Evaluating Radiographs for Equine Foot Management

Pete Healey, APFI

This book is a result of direct and indirect help from a lot of people. I want to thank the staff and veterinarians at Alamo Pintado Equine Medical Center, most importantly Dr. Mark Rick, who always gave me the go-ahead on research projects; Doug and Sue Herthel for their gracious hospitality and letting me work out of their clinic all these years; My longtime friend and client, Jill Brady, who did an excellent job on proof reading; Bill Reynolds for his advice, layout and cover design; Robin Ireland; David Jensen, DVM; Bob Smith APF-1 CF; Amy Rucker, DVM; Richard Mannsman, VMD and Debra Taylor, DVM. Thanks to Karen and Charlie for their loving support.

$1.00 from every book sold will be donated to the Equine Podiatry Education Foundation.

Healey Enterprises, LLC
PO Box 704
Los Olivos, CA 93441
healeyenterprises.com

ISBN-13: #978-0-692-05105-4
Manufactured in the United States of America

For the horses:
Their feet will *tell* us what they need
if we will *listen* to what we see.

TABLE OF CONTENTS

FORWARD

"The common eye sees only the outside of things, and judges by that, but the seeing eye pierces through and reads the heart and the soul, finding there capacities which the outside didn't indicate or promise, and which the other kind couldn't detect."

—Mark Twain

It is an exceedingly rare opportunity as a veterinarian to form an alliance with a farrier who is as driven by the desire of his 'seeing eye' to 'pierce through and read the heart and soul' of an equine hoof in the manner of Pete Healey. Pete has challenged me from our very first collaborative efforts in podiatry to see more than what is presented on the surface and to delve deeper in understanding the functional anatomy and mechanics of the equine hoof.

Pete Healey is committed without condition to understanding the physics of equine podiatry. His interest in the physics or 'knowledge of the nature' of the hoof has propelled him to investigate the static anatomy as well as the hoof's mechanics in motion. He continually studies the effects of trimming and shoeing techniques on break-over distance, moment arms, palmar/plantar angles and effects on the hoof's center of rotation, to name a few of his areas of interest. These concepts and others have become part of the common vocabulary in Pete's communication with farriers, veterinarians and equestrians in his quest to aid the horse.

Pete has always valued the power of a common language for those of us in the arena of equine podiatry. This book provides a starting point or nidus that should motivate us all to continue to investigate these principles of hoof physics and to develop a common language with which to debate these principles.

David Jensen, DVM
2017

INTRODUCTION

"look where everyone else has been looking,
but see what no one else has seen."

—Robert Kennedy

I am a student of the horse's foot, although I haven't always been. I was never that passionate about shoeing horses for a living but I like helping horses and the mechanics of the foot intrigues me.

About 15 years ago, I started shoeing for a large veterinarian clinic that sees a substantial caseload of lame horses. I was getting good results working on some of these horses but I really couldn't answer why. This clinic does not have a big interest in podiatry but it has the best diagnostic equipment available and they have been very generous about letting me utilize the information it provides to develop my knowledge. The great Yogi Berra said, "You can see a lot by just observing." Over the years, I have compared hundreds of radiographs to the live foot of the horse, but there is more to it than that. I have had the great fortune to work with and study the work of Dr. Ric Redden, who has developed the gold standard for measuring soft tissue parameters and bone angles on lateral radiographs. I have also gained important information from the literature of Dr. Christopher Pollitt. Being able to work alongside my good friend Dr. David Jensen has been a tremendous help. Along the way, I have recognized additional parameters, both structural and mechanical, that are relevant to the foot.

There is so much information that can be gleaned from a radiograph; the biomechanical formula of the foot is in front of us, in the numbers and angles. The mechanics of the foot has a tremendous impact on soft tissue, bone, tendon and ligament stresses in the foot, as well as up the limb. When we dial into our knowledge of the impact and meaning of these measurements, our resulting understanding allows us to provide treatment that can seriously influence the soundness and longevity of the horse. The horse has a better chance at avoiding the fate referred to by generations of horsemen as, "No foot, no horse."

Some in the industry feel reading radiographs is best left to the veterinarian and for diagnosing pathology, this is true. When evaluating the foot, the horse is better served if all involved in his care have in-depth knowledge of what is going on inside the hoof capsule, especially the farrier – after all, what's the secret? Everything we do to the outside of the hoof has an impact on the inside, and everything on the inside has an impact on what we can do to the outside.

Owners, trainers, veterinarians and farriers can all have misconceptions about why some feet are the way they are and how to manage them. The more information we can gather, the better the decisions we can make when it comes to the purchase of a horse, his performance prospects or his management requirements. The bottom line is – what is best for the horse?

Pete Healey, APF I
Santa Ynez, CA 2017

ANATOMY OF THE RADIOGRAPH

Dorsopalmar Projection

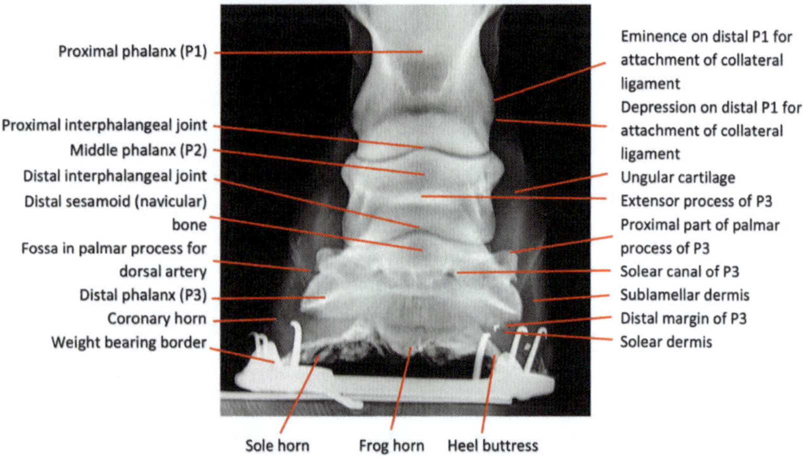

Latero-medial Projection

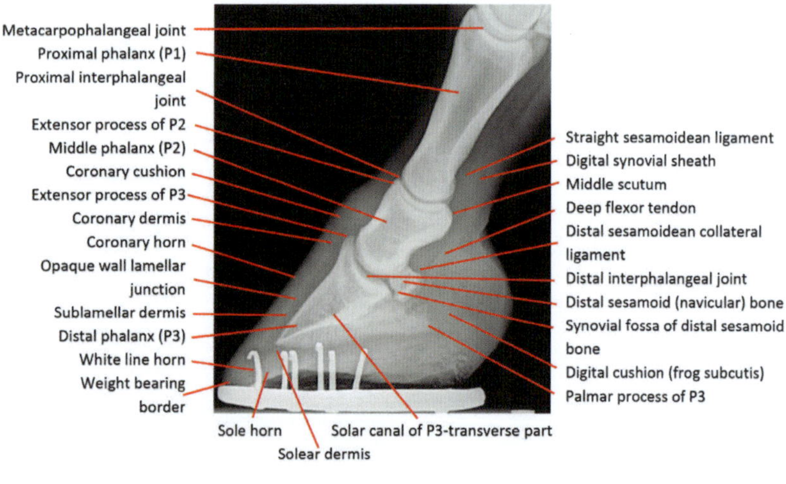

Pollitt C. Radiography of the Foot. In: The Illustrated Horse's Foot: A Comprehensive Guide. Elsevier, St. Louis, Missouri; 2016: 168-173.

1

BASIC UNDERSTANDING OF THE RADIOGRAPH PROCEDURE

There are basically two images that pertain to the soft tissue, balance and mechanics of the foot. These are the lateromedial (LM) and dorsopalmar (DP) projections. For the best evaluation results, the person reviewing the radiographs needs good quality radiographs of properly positioned and marked feet (Eggleston 2012, Redden 2003). Techniques for ensuring high quality radiographs are described in detail elsewhere (Eggleston 2012, Redden 2003, Page et al 2002) but as a farrier, horse owner or veterinarian you should recognize that the procedure is being done correctly.

For the LM and DP projections, both the front or hind feet should be on a block of adequate height whereas the beam of the generator strikes the foot mid sole, half way between the ground bearing surface of the foot and the palmar plane of P3. The cannon bone of the leg should be as perpendicular to the ground as possible. Standing a horse with the leg forward or back will distort the pastern axis and ground mechanics of the foot. The blocks should have a wire embedded in them to delineate the foot surface of the block. This is helpful for determining the ground mechanics of the foot, whether barefoot or shod. When placing the feet on the blocks, the foot getting radiographed should be at the medial edge of the block so the x-ray plate can be placed directly against and parallel to the foot. This eliminates magnification error. Having the plate against the foot is not as critical if the radiograph is to be measured in a calibrated system (a).

For routine LM and PD views, do not pull the shoes; it is best to leave them on. They can provide useful information regarding balance, hoof mass, break-over, sole depth and load zones associated with the current shoeing strategy (Redden, 2003). Because of radiographic "burn-off" of the dorsal crest in the lateral view, the surface of the dorsal wall should be delineated using a radiopaque marker. This is important because there are at least four measurements derived from this marker. Barium paste works best because it tightly hugs the contour of the hoof wall, however a wire marker will also work. Center the marker on the dorsal wall and extend it from the top of the coronary band at the hair line to the distal edge of the foot (Fig. 1.1).

Center the x-ray beam midway between the palmar surface of P3 and the ground surface of the foot, and half way between the toe and the heel (Fig. 1.2). When centered properly, the branches of the shoe, as seen on a shod foot, will be superimposed on each other, as well as the wings of P3 on a mediolateral balanced foot. Centering the beam on the coffin joint is useful for evaluating this area but will produce an oblique LM view that will distort soft tissue measurements (Fig. 1.3). Elevation of the beam may be necessary for a foot with excessive sole depth.

The DP view is commonly taken with the plate behind the foot. The fetlock prohibits the plate from contact with the foot which produces a somewhat magnified view. This is acceptable because most of the evaluation is concerned with mediolateral balance of P3 and coffin joint tilt. Like the LM views, using a calibrated system can give more accurate distance lengths.

It is important to distinguish the left foot from the right in presentation of the radiograph. Usually a marker is taped or clipped to the x-ray plate to designate left or right. This can be a L or R marker or a coin or something radiopaque. On the LM view, this is positioned dorsally on the toe side of the plate. On the DP view, it is positioned on the lateral side of the plate. Lateral radiographs of the left foot should be presented with the toe facing left. Radiographs of the right foot should be presented with the toe facing right. This makes clear which foot is being presented.

Although some people are still developing traditional cassettes, most are using digital radiology which produces an image that can be measured on a computer screen with a mouse. These can be emailed or printed out and measured by hand. To ensure that the image is the accurate size, tape a metal marker of determined length to the x-ray plate and then measure the print-out to determine adjustments to the measurements.

The formula for this is:

(Actual length of marker ÷ Radiograph length) Length measured

Radiographs can also be evaluated using a percentage of the Palmar cortex measurement of P3 which is described in Chapter 13.

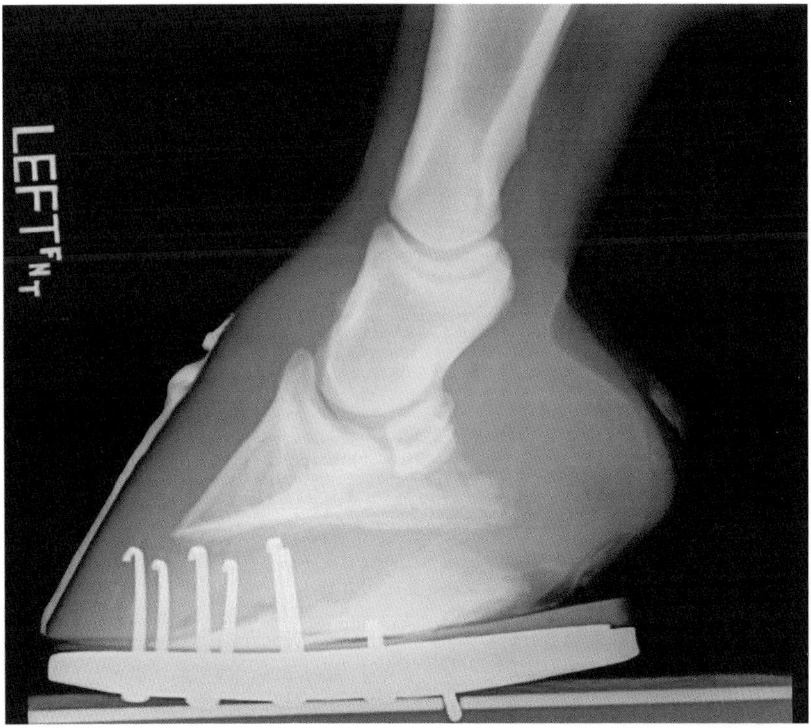

Fig. 1.1: Foot with a barium dorsal wall marker.

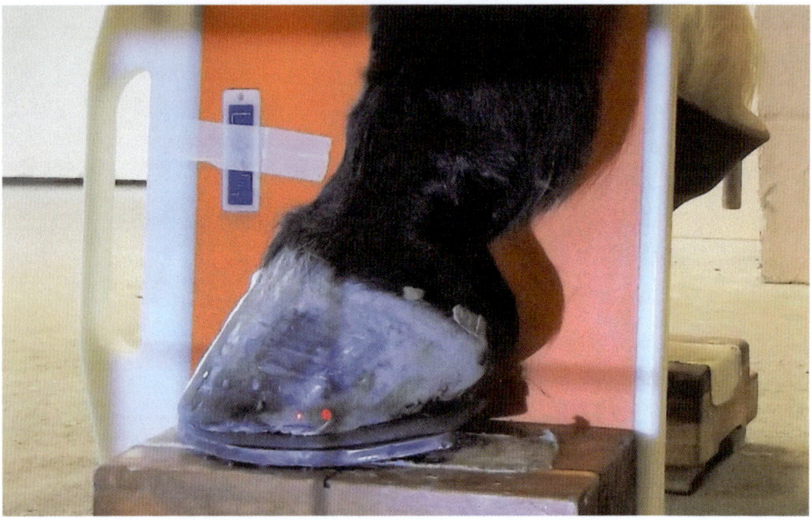

Fig. 1.2: Position of foot for a mid-sole lateral projection.

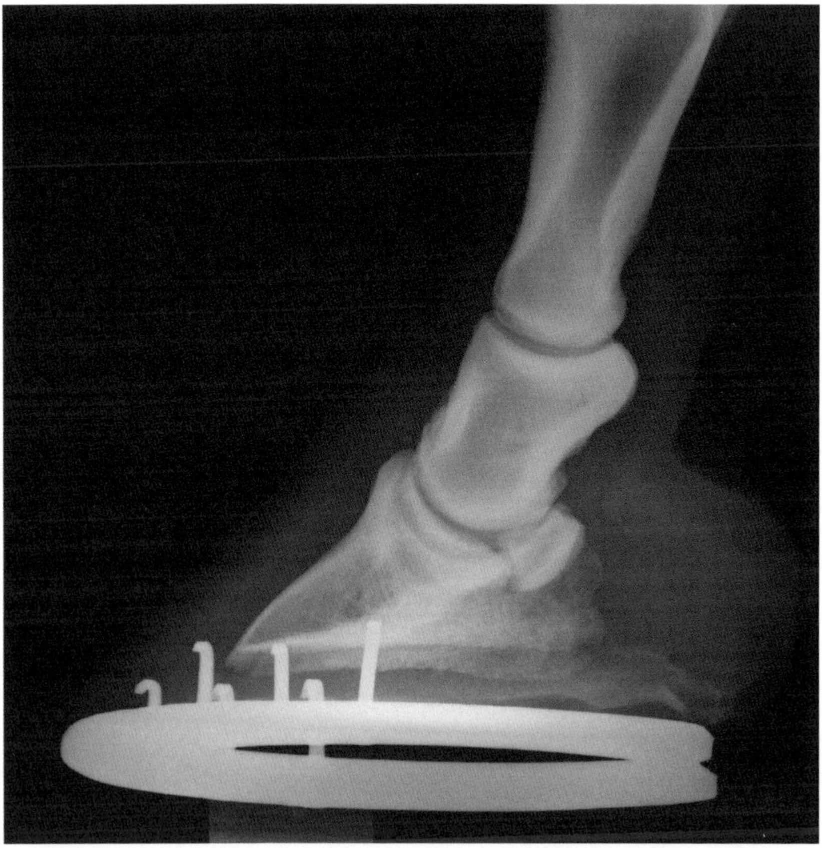

Fig. 1.3: An oblique lateral; this is fine if the area of interest is the coffin joint but not useful for measuring soft tissue.

Glossary

Dorsopalmar: Front to back

Balance: An even or appropriate distribution of weight, foot mass and moment arms around center of rotation, also referred to as *equilibrium*.

Lateral: Outside, away from the body

Lateromedial: Outside to inside

Mechanics: The branch of physics that deals with motion

Medial: Inside, closer to the body

P3: Coffin Bone, Distal Phalanx

References

Eggleston, R.B., Equine Imaging: The Framework for Applying Therapeutic Farriery. Vet Clin Equine 28 (2012) 293-312.

Page B.T., Hagen T.L., Break-over of the hoof and its effect on structures and forces within the foot. Journal of Equine Vet Science 22:6 (2002) 258-264.

Redden, R.F., Radiographic imaging of the equine foot. Vet Clin Equine 19 (2003) 379-392.

A Metron Equine Hoof Conformation Software System, Epona Tech, P.O. Box 361, Creston CA 93432

2

UNDERSTANDING THE BASIC MECHANICS OF THE FOOT

The purpose of the radiograph is to evaluate the internal structures of the foot to determine its function and health. Physics is a science that deals with matter and energy and their interaction. Detailed descriptions of biomechanics and movement are described elsewhere (Parks 2003, Rooney 2008). Effective management decisions regarding trimming and shoeing rely on understanding basic physics. It is important to understand the impact of distortion and compression on the hoof.

As the horse bears weight on the leg, the fetlock descends to the ground, dorsiflexing the metacarpophalangeal joint (Fig. 2.1). This moment (turning force) is countered by tension in the superficial flexor tendon and suspensory ligament. This produces a moment about the coffin joint that is countered by tension in the deep flexor tendon (DFT), (Parks 2003) which produces an equal tension in the extensor tendon. Center of Rotation (COR) is located in the center of the condyle of P2 (Colles 1983). COR is like an axle on a wheel and it is around this point that rotation of P3 occurs. Conformation of the foot and leg influence how weight is distributed about COR, as well as the moment arms (MA), which affect the ground surface of the foot.

A vector has direction and force and every force has an equal and opposing force. This principle underlies the role of the DFT. The tendon's amount of lift or lack of suspension, combined with the ground mechanics, influence load on the foot. When looking at the diagram in Figure 2.1, note the yellow line that is the line of action (LOA) of the extensor and deep tendons. The distance between this line and the red line of COR, at the ground surface of the foot, creates a MA (Wilson et al 2011). This MA is a product of the Palmar Angle (PA) of the foot and the MA of the Ground Reaction Force (GRF) dorsal to COR. The distance between these two lines has a direct bearing on the amount of load placed on the heels palmar to the vector of COR at P3 (dashed yellow line). If P3 rotates whereas the yellow line converges with the red line at the ground surface of the foot, there would be no MA about the coffin joint in the static foot.

The length of the MA of the GRF dorsal to COR is determined by the radius on the ground surface of the foot or shoe. This has a direct bearing on leverage induced distortion to the dorsal complex of the foot. How much pressure is produced at the apex of P3 is determined by the pull of the DFT and the amount of GRF at the toe.

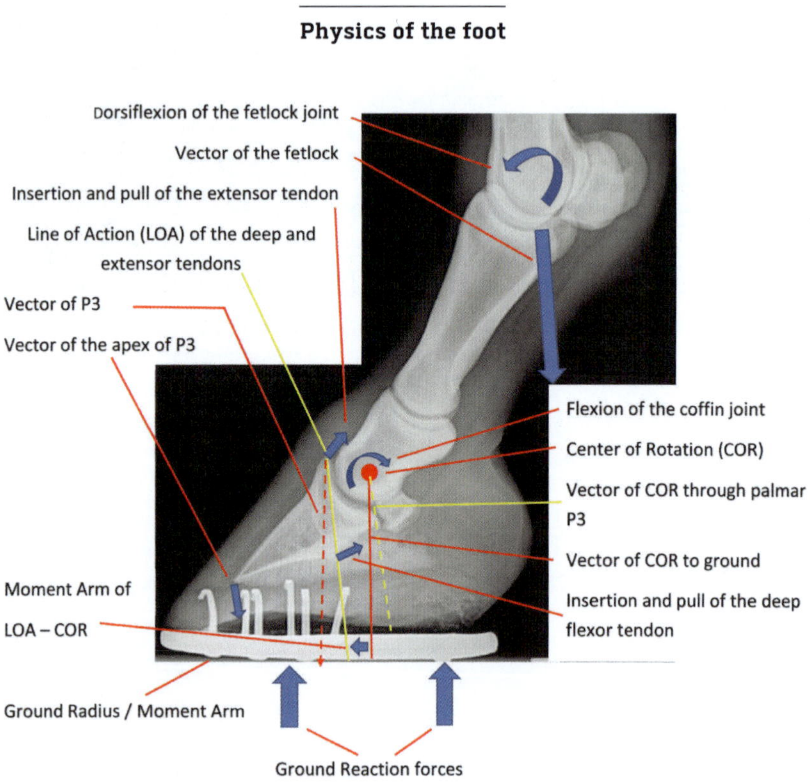

Fig. 2.1: Biomechanics of the equine foot.

Glossary

Center of Rotation: Located in the center of the condyle of P2; this is the biomechanical pivot point for weight distribution proximally and moment arms distally

Dorsal: Toward the front

Dorsiflexion: A backward bending, toward the front

Flexion: The action of bending around a joint, extension

Ground Reaction Force: The force exerted by the ground on a body in contact with it

Moment: A turning force

Moment Arm: The length between a joint axis and the line of force acting on that joint

Palmar: Toward the back, underneath

Vector: Direction and force

References

Colles C. Interpreting radiographs 1. The foot. Equine Vet J. 1983; 15: 297-303.

Parks A., Form and function of the equine digit. Vet Clin Equine 19, Saunders, Philadelphia (2003) 285-387.

Rooney J.R., Functional Anatomy of the Foot. In: Floy A., Mansmann R.A., ed. Equine Podiatry, St. Louis Missouri: Saunders Elsevier, 2008; 57-73.

Wilson A., Weller R. The Biomechanics of the Equine Limb and Its Effect on Lameness. In: Ross M.W., Dyson S.J., ed. Diagnosis and Management of Lameness. Saunders Elsevier, St. Louis Missouri; 2011: 270-281.

3

FIRST VIEW IMPRESSION

Your first evaluation of a radiograph is with your eyes. What do you see? Is it just an image or is it a set of tissue and bones that perform a function (an organ) (Fig. 3.1)?

The foot is an integrated organ (Redden 2003) where one mechanical aspect affects another aspect. This in turn has an effect on soft tissue and bone.

Normal and abnormal variations can be seen but they can only truly be evaluated through measurement. The good news is that the act of measuring lots of x-rays over time will bring a higher focus to your radiographic eye. You will see and understand far more than your first glance. It takes more than a clear image to maximize the use of a radiograph; a practiced, measured and informed understanding make the radiograph a powerful diagnostic tool.

Overall Image Assessment Lateral Radiograph

1. Is the lateral flat and straight? Are the bulbs of the heel and the branches of the shoe superimposed on each other?
2. If the image is flat, are the wings of the coffin bone superimposed on each other or is P3 oblique showing two wings?
3. Is the palmar angle of P3 negative, flat or positive?
4. Is P3 balanced in the hoof capsule? Is the dorsal face of P3 parallel to the external hoof? Are the hoof-lamella zones equal to the sole depth?
5. What is the mechanics of the toe? Is there a radiolucency between the weight bearing surface of the toe and the block or is the toe buried into the block? How much toe is out in front of the apex of P3?
6. What is the posture of the pastern? Does it have a nice slope or is it too vertical or too extended?
7. Are the bones normal or do they have irregularities, blemishes, fractures or abnormal bony growths?

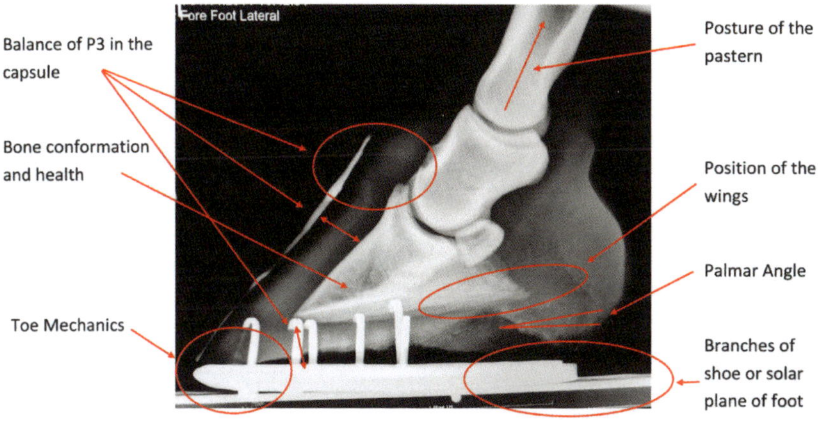

Overall Image Assessment

Fig. 3.1: The first view of the radiograph offers up information as to the quality of the radiograph and what areas of the foot bring immediate attention or concern.

References

Redden R. F., Hoof capsule distortion: understanding the mechanisms as a basis for rational management. Vet Clin Equine North Am Pract 19, Saunders, Philadelphia (2003); 443-462.

4

SOFT TISSUE MEASUREMENTS

Why measure? Because if we measure, we can monitor and manage. I think the reason a lot of people don't measure their radiographs is because they don't know what information is important and they think it is too time consuming. Once you learn the procedure, measuring becomes very easy and opens your mind up to a lot of different aspects of the foot. New skills take time to learn but they are worth the investment. Measuring will increase your accuracy over failure to measure.

It is best to measure in the metric system in millimeters. Using the American system of inches and fractions is very confusing. The metric system is a straight scale of numbers; it is much easier to work with and is the universal system in use by most veterinarians, and in common use in book descriptions and review papers as well. Once you learn the parameters, you will be able to visualize the lengths very readily on the foot.

Coronary Band – Extensor Process Distance (CE)

The coronary band-extensor process (CE) measurement is the vertical distance between the top of the dorsal wall marker, which is placed at the proximal limit of the hoof wall at the hair line, and the proximal limit of the extensor process (Fig. 4.1). Marking the coronet band this way is more consistent and easier than another method which describes the proximal limit as the transition between soft and hard horn (Cripts, et al 1999). The CE measurement can vary in depth from horse to horse and foot to foot. One author noted that in most normal horses, the CE distance is in the range of 0 to 15 mm (Redden 2008). In a study of 138 non-laminitis front feet by this author, the CE ranged from 4 to 26 mm with a mean CE of 12 mm.

This measurement can vary in depth between left and right feet as dictated by the amount of suspension of the deep flexor tendon-muscle unit.

In mismatched feet, the bigger, wider foot usually has an increased CE distance. This lower angled limb will usually coincide with a more developed proximal limb (shoulder) as viewed from the front of the horse and will usually have a lower palmar angle than the opposite more upright foot.

In a non-laminitis foot, a CE distance that is greater in one foot may coincide with a wider than normal proximal hoof-lamella zone measurement and a dorsal bone angle of P3 that is negative to the proximal hoof angle. This could be an indication of a caudal displacement in the hoof capsule. The CE distance can also be an indicator of P3 displacement during a laminitis event. Following this measurement in conjunction with the hoof-lamella zone and dorsal sole thickness in subsequent radiographs can determine further sinking or recovery of the foot with the current mechanical therapy.

Hoof – Lamella Zone (HL)

The hoof-lamella (HL) zone is the perpendicular distance between the dorsal surface of P3 and the outer surface of the dorsal hoof, as measured to inside the dorsal wall marker. A wall marker of Barium Paste is best because it hugs the surface of the dorsal wall creating a more accurate measurement. The marker is important due to the fact that edge burn-through (saturation artifact) at the periphery of soft tissue is a common artifact with digital radiographs which can result in inaccurate assessment of the hoof wall thickness (Jimenez, et al 2008).

The HL is measured in two places: proximally just below the extensor process, and distally near the tip of P3 (Fig. 4.1). In a normal foot, the lamellar corium and the dorsal hoof wall are the same thickness. In a foot with a HL zone of 15 mm, each zone measures 7.5 mm (Redden 2008).

A healthy HL measurement usually corresponds to 25% of the measurement of the palmar cortex of P3 (see Palmar Cortex measurement) (Pollitt 2001). Normal HL zones can vary from 15 to 16 mm in a light horse to 20 mm in Standardbreds, normal HL in Warmbloods depends on the size of the foot (Redden 2008). This relates again to the palmar cortex length of P3.

A normal proximal measurement with a wider distal measurement indicates distal leveraging of the hoof wall. A normal proximal measurement with a narrowing distal measurement can indicate ground abrasion or over-rasping of the distal dorsal wall – note the thickness between the two zones. A wider than normal proximal measurement combined with a negative dorsal bone angle to hoof angle may indicate stress in the proximal lamina producing a caudal displacement of P3. This can sometimes be identified on the hoof as a bulging dorsal wall (Fig. 4.2). Wider than normal proximal and distal HL zones are indicative of feet with acute or chronic laminitis and usually coincide with other abnormal soft tissue and bone angle parameters.

When widening of the HL zone is present, evaluation of each component is important as it can provide valuable diagnostic and prognostic information (Redden 2008). The lamellar zone widens in cases of laminitis, whereas the horn widens in cases of white line disease.

The lamellar side of the HL zone is a very complex structure. To put into perspective, a normal foot with a HL zone of 15 mm has 7.5 mm of horn and 7.5 mm of lamellar corium. A 1 mm increase on the lamellar side of this structure is an extra 13%. This swelling or stretching can account for lack of blood profusion, pain and structural instability.

Sole Depth (SD)

Sole depth (SD) is the distance measured from the solar margin at the apex of P3 to the radiolucent border of the sole. This is usually measured in a perpendicular line to the palmar plane of P3 (Fig. 4.1). Healthy feet on a light breed horse should have a minimum sole depth of 15 to 18 mm (Redden 1993 a). Sole depth is deemed adequate when it approximates the HL zone distance. In other words, a 16 mm HL zone requires approximately 16 mm of sole depth for proper functioning (Redden 2003 a).

A sole depth of less than 15 mm is clinically significant. Venograms in horses with a sole depth less than 15 mm show solar papillae that are bent, compressed or absent (Redden 2008). Some horses lack epidermal sole mass, most notably horses that are in wet conditions or on permanent pasture. When evaluating the distance under the apex of P3 on these feet, it may be necessary to measure through the radiolucent cup of the sole to the block or the foot surface of the shoe to consider ventral depth of the foot (Fig. 4.1).

Heel Depth (HD)

Heel depth (HD) is defined as the distance from the ventral surface of the palmar aspect of the wing of P3 to the ground surface of the foot (Fig. 4.1). An appropriate HD measurement should be equal to the SD distance plus additional hoof mass. This provides adequate palmar angle to sustain the hoof angle and a properly positioned hoof-pastern axis.

Venogram studies suggest that a 3 to 5 degree palmar angle is needed to sustain the sole and other sensitive structures of the heel apertures (Redden 1993 a).

An adequate heel depth is a product of a healthy digital cushion-frog unit. Most feet that have less than 20 mm of ventral depth under the palmar process of P3 have collapsed heels and often show wall buckling in the quarter-heel region. It is often recommended to maintain at least 20 mm of depth for these reasons (Redden 2003 b) (Morrison 2004).

Palmar Cortex (PC)

The palmar cortex (PC) of P3 is not a soft tissue measurement but it relates to soft tissue. This is a measurement of the distance from the tip of P3 to the articulation of P3 with the navicular bone (Fig. 4.3). This measurement provides a relative scale when evaluating the soft tissue and relative mass of the foot. Research shows a healthy HL zone corresponds to 25% of the measurement of the PC (Pollitt 2001). In a study of 138 feet by this author, 25% of the PC measurement equaled + - 1 mm the distance of the proximal HL zone. This did not take into consideration that many of these feet presented a caudal displacement in the hoof capsule.

The usual range of the PC length is from + - 55 mm in a light breed horse to + - 80 mm in a warmblood horse. This gives perspective to the size of P3 within the hoof capsule. For example: an average 1300-pound warmblood horse that should have a PC measurement of 75 mm is radiographed with a PC measurement of 60 mm which shows up as a congenitally small P3.

When comparing several radiographs of the same foot, the PC measurement can be useful as a marker in determining if magnification error is present. It is handy to have a chart that has 25% of the PC distance posted for easy reference when measuring radiographs. It allows efficient evaluation of the HL, SD, and HD measurements (Fig. 4.4).

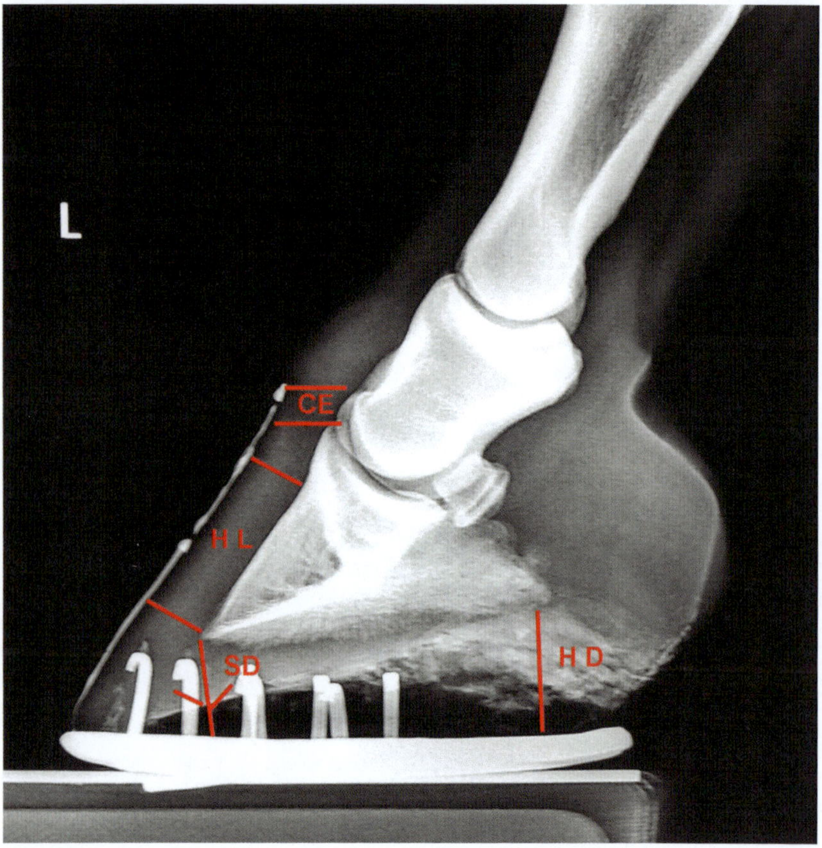

Fig.4.1: Parameters for measuring CE, HL, SD and HD distances.

Fig. 4.2: The bullnose shape of a foot with a caudal displacement of P3. This foot has been excessively rasped to eliminate dorsal and mediolateral flaring. The distortion of the foot is a product of laminar leveraging from ground reaction forces.

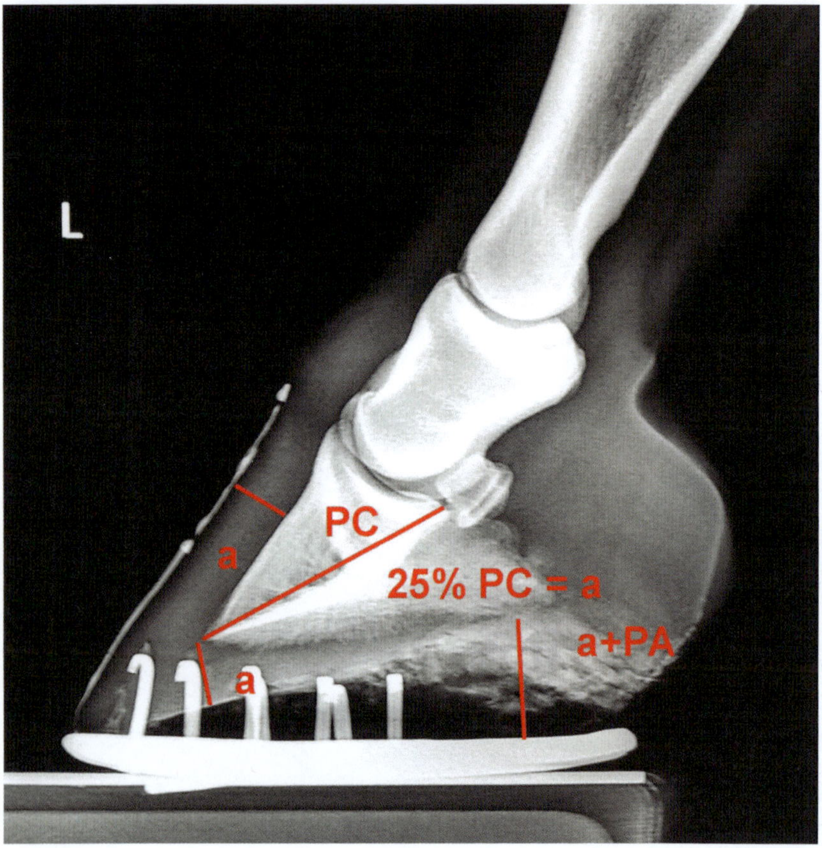

Fig. 4.3: The PC measurement is from the apex of P3 to its articulation at the center of the navicular bone.

Quick Reference Chart for the Palmar Cortex Measurement and Approximate 25% Hoof-Lamella and Sole Depth Value

PC	HL/SD
52	13
56	14
60	15
64	16
68	17
72	18
76	19

Fig. 4.4: The reference chart gives an approximate value of 25% of the PC in millimeters.

Glossary

Acute: Of abrupt onset

Chronic: A persisting condition

Distal: Away from the center of the body, on the leg this would be the lower end

Epidermal: External layer of horn wall or sole comprised of dead cells made by the dermal layer

Papillae: A vascular protuberance of the dermal layer of the wall horn or sole extending into the dermal layer

Proximal: Closer to the center of the body, on the leg this would be the upper end

Venogram: Radiograph of a foot after the injection of a radiopaque contrast which fills the arteries and veins

Ventral: Relating to the underside

References

Cripts P.J., Eustace R.A. Radiological measurements from the feet of normal horses with relevance to laminitis. Equine Vet J. 1999, 31:5, 427-432.

Jimenez D.A., Ambrust L.J., O'Brien R.T., et al. Artifacts in digital radiography. Vet Radiol Ultrasound 2008; 49: 321-332.

Morrison S. Foot Management. Clinical techniques in equine practice, Elsevier 2004; 71-82.

Redden R.F. The use of venograms as a diagnostic tool. In Proceedings: 7th Bluegrass Laminitis Symposium 1993; 1-6.

Redden R.F. Hoof capsule distortion: Understanding the mechanisms as a basis for rational management. Vet Clin North Am Pract 2003 a: 19; 443-462.

Redden R.F. Clinical and radiographical examination of the equine foot. In Proceedings: 49th AAEP convention 2003 b; 169-178.

Redden R.F. Radiographic Views of Most Value to Farriers, In: Floyd A., Mansmann R.A., ed. Equine Podiatry, St. Louis, Missouri, Saunders Elsevier,2008; 199-202.

Pollitt C. The Horse's Foot. In; Equine Laminitis, 1st ed. Canberra: RIRCC Publications 2001; 4-14.

5

ANGLES RELEVANT TO P3 AND THE HOOF

Palmar/Plantar Angle (PA)

Known as the palmar angle (PA) in the front foot and the plantar angle (PA) in the hind foot, the PA is the angle produced by the palmar surface of P3 in relation to the ground (Fig. 5.1). The angle formed by the wing of P3 is the most accurate anatomical point of reference as some feet will have a remodeled aspect to the apex of P3. The PA has been noted to be from 2 to 10° in a sound horse (Van Heel et al 2004). It has been published that the PA is ideally 3-5° greater than the dorsal angle of P3; this allows P3 to sink in a distopalmar direction during weight bearing from the flexion of the coffin joint and the dorsiflexion of the fetlock, which produce compression on the caudal hoof (Denoix 1999). This angle range may not take into consideration a foot with a P3 bone angle of less than 50° that would require more PA to achieve a balanced hoof angle.

The PA is relative to the amount of digital cushion, sensitive and horny frog in the palmar foot (Mansmann et al 2007) and a positive PA usually ensures a healthy digital cushion – frog support system. The PA is largely regulated by the tension of the deep flexor tendon and the mechanics of the foot dorsal to center of rotation (COR). In one study, the average foot lost 2° of PA in six weeks due to grow-out (Van Heel et al 2005).

The PA has a direct relation to the hoof-pastern Axis (HPA) as a 2° increase or decrease of the PA will change the extension of the HPA 5° (Healey 2016).

Bone Angle (BA)

The bone angle (BA) is the angle made by the dorsal and ventral surfaces of P3 (Fig. 5.2). On a radiograph that displays bone remolding or erosion to the apex of P3, using the proximal aspect on the dorsal surface and the angle of the wings is best for an accurate measurement.

The BA measurement is critical in understanding the target hoof angle (HA) for a particular foot. A good conformation foot has a BA of +/- 50°, this combined with a 3-5° PA equals a HA of 53-55° which the industry loves and will usually create a correct HPA (0° axis when standing weight bearing on that limb). A foot with a BA of 40° requires considerable PA to achieve normal standards. This is important when discussing foot conformation between owners, veterinarians and farriers.

The low BA foot that has adequate digital cushion and frog can have a different conformation of the palmar foot than a normal BA foot. The higher PA needed to offset the low BA creates a foot with narrower heels that grow in a more dorsal position. The industry philosophy of trimming the heels back to the widest part of the frog doesn't work well with these feet as it will upset the balance with the PA and the HPA. BAs can vary between the left and right foot on the same horse.

The BA of every foot should be a consideration of any horse during a pre-purchase exam as a foot with a BA in the low 40° range may require a lifetime of wedge shoes or pads to stay physiologically correct and sound. In a study by this author of 138 feet from 69 horses that had been diagnosed with pathology from an MRI examination, the average BA of these feet was 45.2°.

Hoof Angle (HA)

The hoof angle (HA) is the angle produced by the dorsal hoof wall relative to the ground surface of the foot (Fig. 5.3). The proximal half of the dorsal wall is the most accurate to measure, as often the distal half has been manipulated by rasping or ground reaction force leverage. There is general agreement that the dorsal HA should match or nearly match the dorsal pastern angle (Bach et al 1995). The data describing recommended hoof angles in the literature is quite variable (Avisar 2007).

The HA is best evaluated when used in conjunction with the bone and palmar angles:

BA + PA should = HA

But we need to take into consideration its relationship to the dorsal bone angle of P3.

Dorsal Bone Angle (DBA)

The dorsal bone angle (DBA) is the angle made by the dorsal surface of P3 and the ground surface (Fig. 5.3). A DBA that is less than the proximal HA can be an indication of caudal displacement of P3. These feet usually show a wider than normal HL zone. This is apparent in the dermal side of the proximal HL zone, which is wider than the distal zone. A venogram study of a foot with a negative DBA to HA (Fig. 5.4) shows lack of vascular profusion to the proximal lamina (Fig. 5.5). This lack of profusion along with the ectodermal stretching of the lamina suggests lack of structural strength of the proximal foot. Another indication of this may be evident in a long CE measurement.

Visual evidence of this may be seen as a bulging horn wall (Fig. 4.2) distal to the coronet band and prominent compression rings in the hoof. These feet may or may not have negative PAs but usually have a craniocaudal imbalance of the foot. A negative DBA will give an inaccurate HPA evaluation. To adjust the foot using the radiograph: take the angle difference between the HA and DBA, divide by 5 and multiply the sum by 2. This equals the amount of angle to add to the PA for correct

HPA evaluation (additional adjustment to the HPA may be needed as indicated by the external measurement of the HPA using a goniometer, (*see Chapter 7*).

(a/5) 2 = PA increase

Balancing out the Angles

Measuring angles on the coffin bone can be somewhat subjective due to the irregular shape of some bones. If the BA + PA does not equal the DBA then one or more angles are off and need to be remeasured. As stated above the DBA does not always have the same angle as the proximal HA, although it may have the same angle as the distal HA.

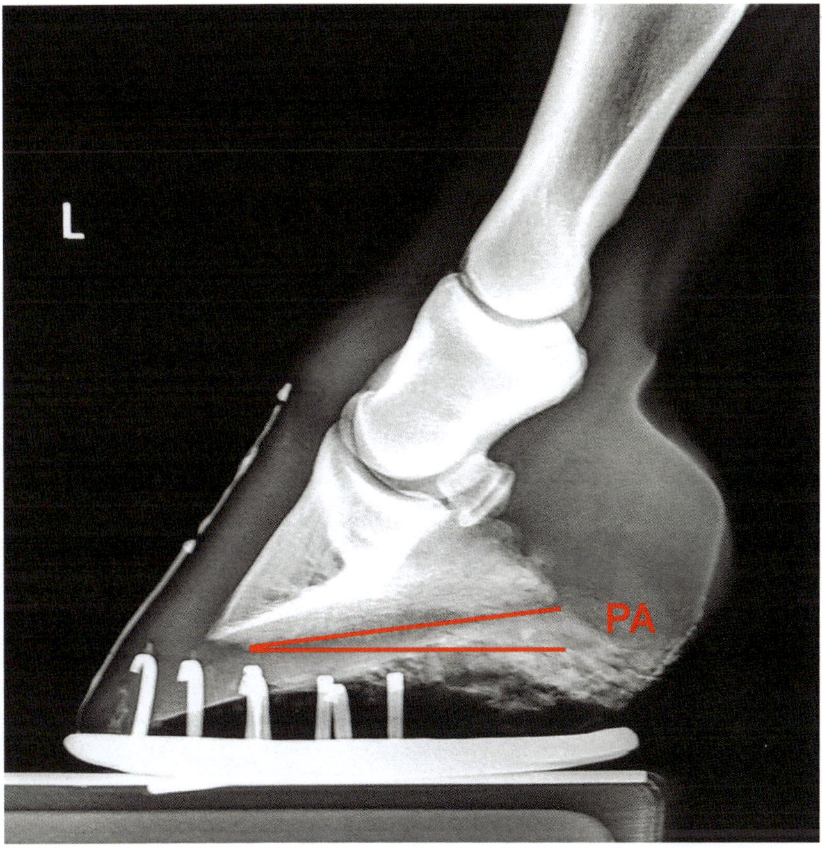

Fig. 5.1: The PA is measured along the ventral surface of P3 using the angle of the wings as a guide.

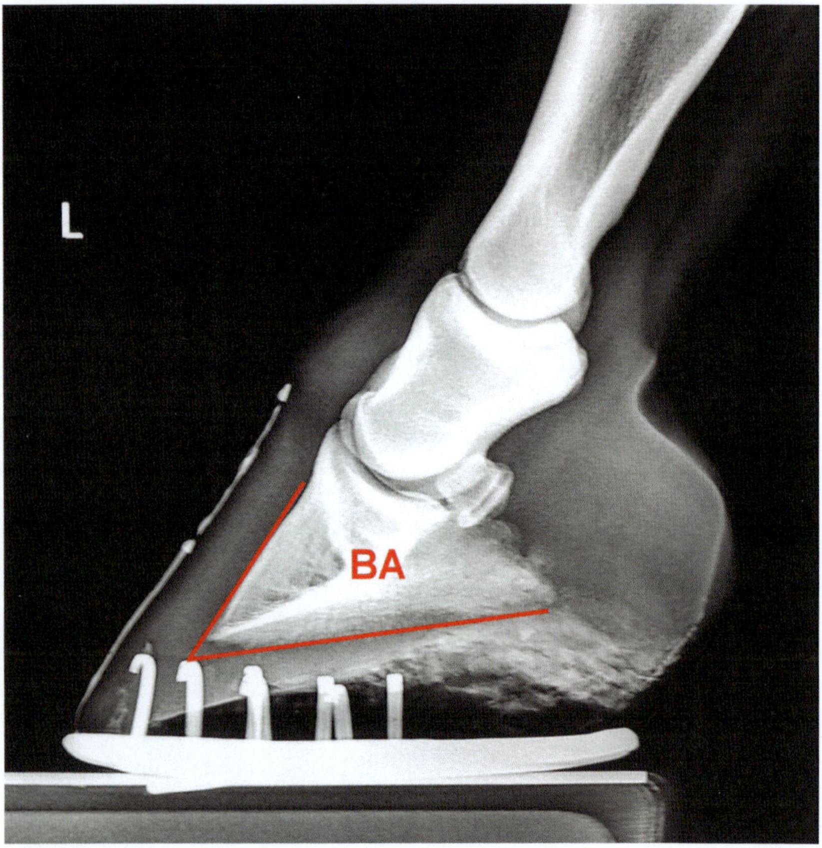

Fig. 5.2: The BA is measured along the dorsal and ventral surfaces of P3.

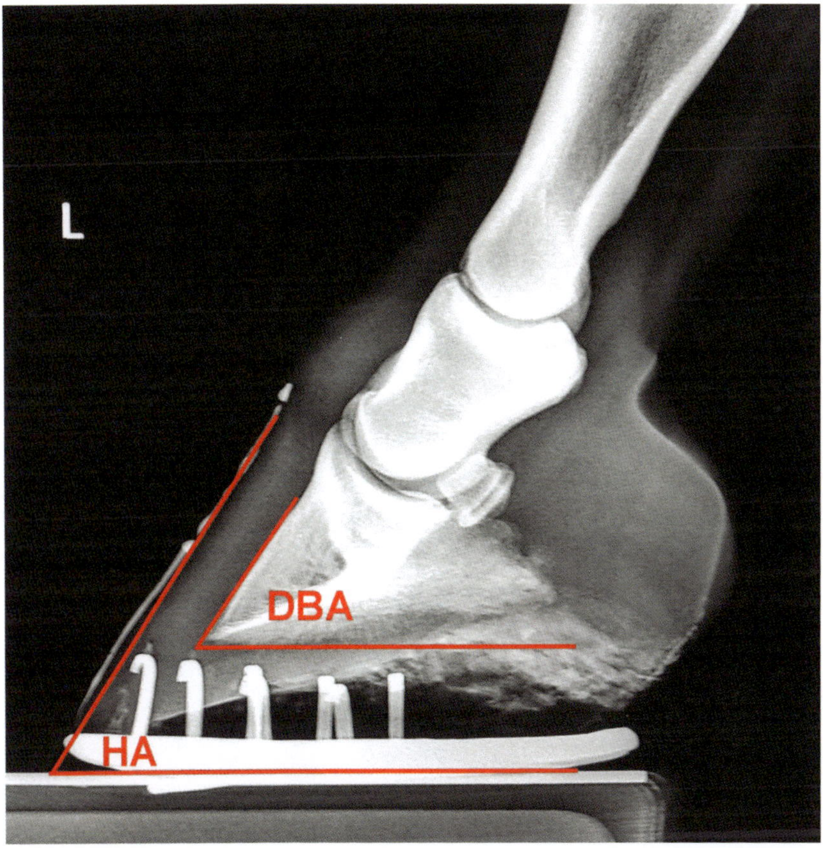

Fig. 5.3: The HA is measured as a line from the dorsal aspect of the proximal hoof to the ground. The DBA is measured as a line from the dorsal aspect of P3 to the ground.

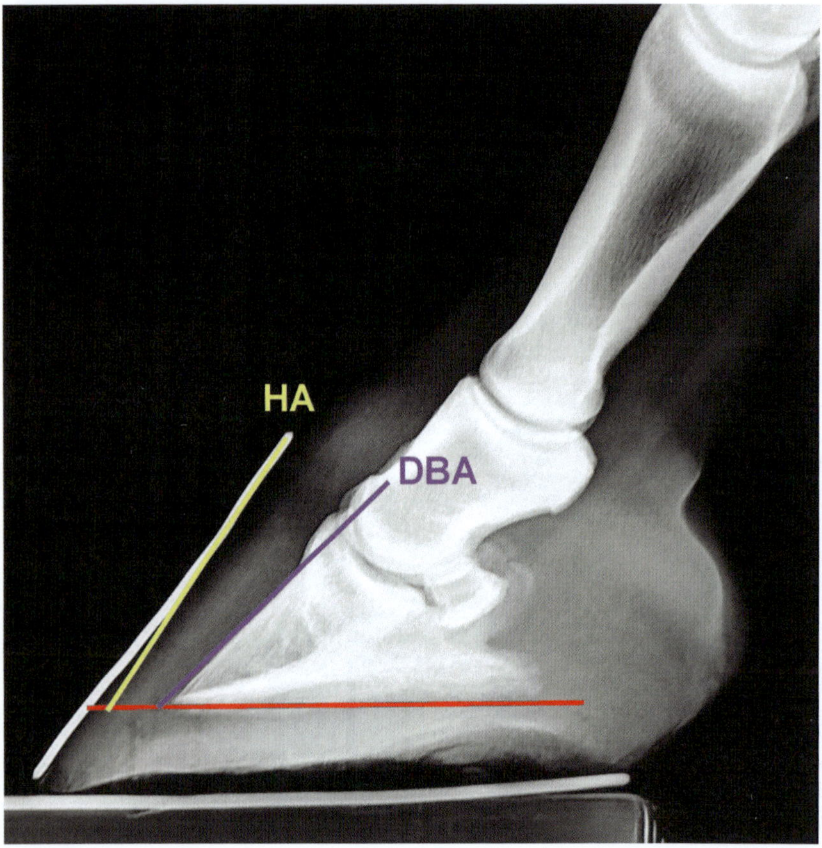

Fig. 5.4: A foot with a caudal displacement of P3; the DBA is at a negative angle to the proximal HA. Note the distance from the top of the extensor process to the top of the dorsal wall marker at the coronet band (CE).

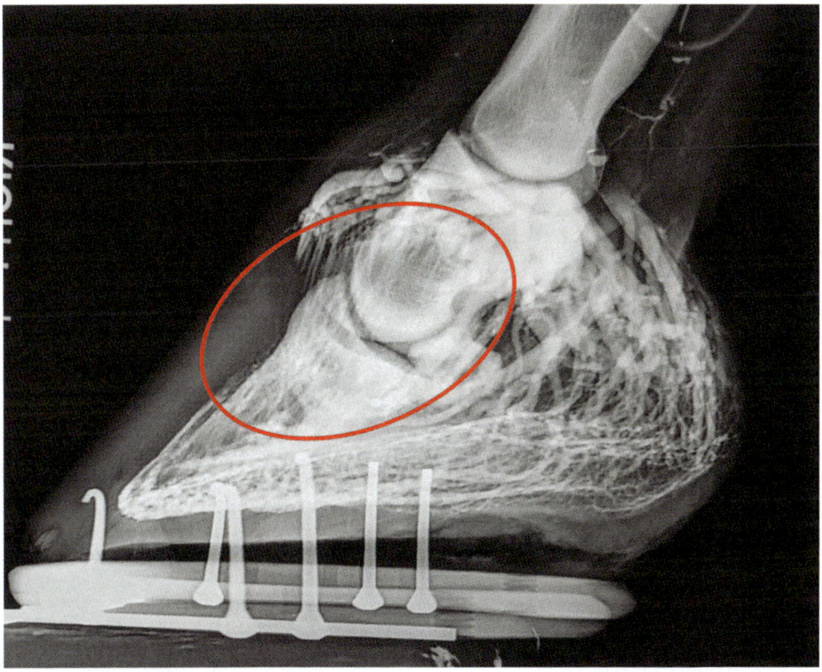

Fig. 5.5: Venogram of the foot in Fig. 5.4; note the lack of contrast in the proximal sub lamellar vessels and coronary plexus.

Glossary

Craniocaudal: Referring to the balance of the angle plane of the foot, this includes the PA and the HPA.

Distopalmar: Directional, down and back.

Ectodermal: Outer layer of the dermis

References

Avisar Y. Hoof Conformation. In: Therapeutic Farriery, A manual for Veterinarians and Farriers, Hippus Publishing, San Luis Obispo, CA 2007; 21-22.

Bach O., Butler D., White K., et al. Hoof Balance and Lameness: Improper toe length, hoof angle, and mediolateral balance. Compend Contin Educ Pract Vet 1995; 17: 1275-1282.

Denoix J.M. Functional anatomy of the equine interphalangeal joints. In Proceedings: Am Assoc Equine Pract 1999; 45:174-177.

Healey P. Measurement of the Hoof-Pastern Axis for Foot Management. In: American Farriers Journal, Lessiter Media, Brookfield, Wisconsin, December 2016; 50-55.

Mansmann R.A., Von Orde K.E. Preventative foot care program. In: Equine Podiatry, Floyd A., Mansmann R. ed. Saunders 2007; 414-431.

Van Heel M.C., Barneveld A., van Weeren P.R., et al. Dynamic pressure measurements for the detailed study of hoof balance: the effect of trimming. Equine Vet J 2004; 36: 778-782.

Van Heel M.C., Moleman M., Barneveld A., Van Weeren P.R., Back W. Changes in location of center of pressure and hoof enrollment pattern in relation to an 8-week shoeing interval in the horse. Equine Vet J. 2005; 37: 536-540

6

ASPECTS OF CENTER OF ROTATION AND THE BREAK-OVER DISTANCE

Center of Rotation (COR)

Center of Rotation (COR) is also known as the Center of Articulation and is located in the center of the condyle of P2 (Colles 1983). There are two aspects to COR.

The force exerted through the skeletal system is acting through COR (Parks 2012) and it has been published that in an ideal foot, a vertical line drawn from COR to the ground should bisect the bearing surface of the foot (Colles 1983) (Fig. 6.1). This theoretically should coincide with a line drawn across the solar surface of the foot through the middle third of the frog and should equal the widest part of the foot (O'Grady et al 2011).

Moment arms around COR determine weight and leverage on different parts of the foot. It has been noted that the distance from the line of action (LOA) of the tendons to COR of a joint creates a moment arm (MA) (Wilson et al 2011). There are basically two MAs in the foot: the LOA of the extensor and deep flexor tendons and the mechanics of the foot dorsal to COR (Fig. 6.2).

Center of Rotation – P3 Distance (COR-P3)

The COR-P3 distance is defined as the distance from a line drawn from COR which is 90° to the ventral plane of P3, to a parallel line drawn from the tip of P3 (Fig. 6.3). This distance can range from 40 +/- mm in a pony to 70 +/- mm in a Warmblood. This measurement can be transferred to the foot to find an accurate estimate of the COR-P3 vector and the dorsal boundary of P3. On the solar aspect of the foot, the COR-P3 vector is the COR-P3 distance that was measured on the radiograph. It is the distance from the palmar junction of the central frog sulcus at the heel bulb to a point just in front of the dorsal limit of the central frog sulcus. The dorsal limit of P3 is an equal measurement from the COR-P3 vector dorsally (Fig. 6.4).

Knowing the dorsal boundary of P3 is important when discussing foot conformation and evaluating the mechanics of break-over as it relates to bone and soft tissue.

Break- Over Distance (BO)

The break-over (BO) distance is the distance from a vertical line drawn at the tip of P3 and 90° to the ventral plane of P3 to the point of break-over (Fig. 6.5). Break-over is the most dorsal location of the solar aspect of the hoof or shoe that contacts the ground (Page et al 2002).

Determining break-over on the radiograph is usually denoted by the visual radius of the foot or shoe along with a radiolucent area between the ground surface of the foot and the x-ray block. The exact point of break-over can be somewhat ambiguous as dictated by the position of the foot on the block (Fig. 6.6). A foot with a flat shoe and in a cranial rotation is easy to read as the toe will be "bottomed out" into the block. Whereas, a foot in a cranial rotation in a rocker type shoe could show break-over at different positions along the radius of the shoe, depending on the position of the belly of the radius to COR and the influence of the deep flexor tendon. A foot with a flat shoe that is in a caudal rotation may show a radiolucent area between the front of the foot and the block, even though the mechanical break-over of the foot is several millimeters dorsal to the tip of P3.

The mechanical interpretation of the toe is important because it has a considerable effect on other parts of the foot. Likewise, other parts of the foot have an effect on it. The longer the break-over distance, the greater the tension on the lamina attachments and the more the horn tubules are bent (Redden 2003).

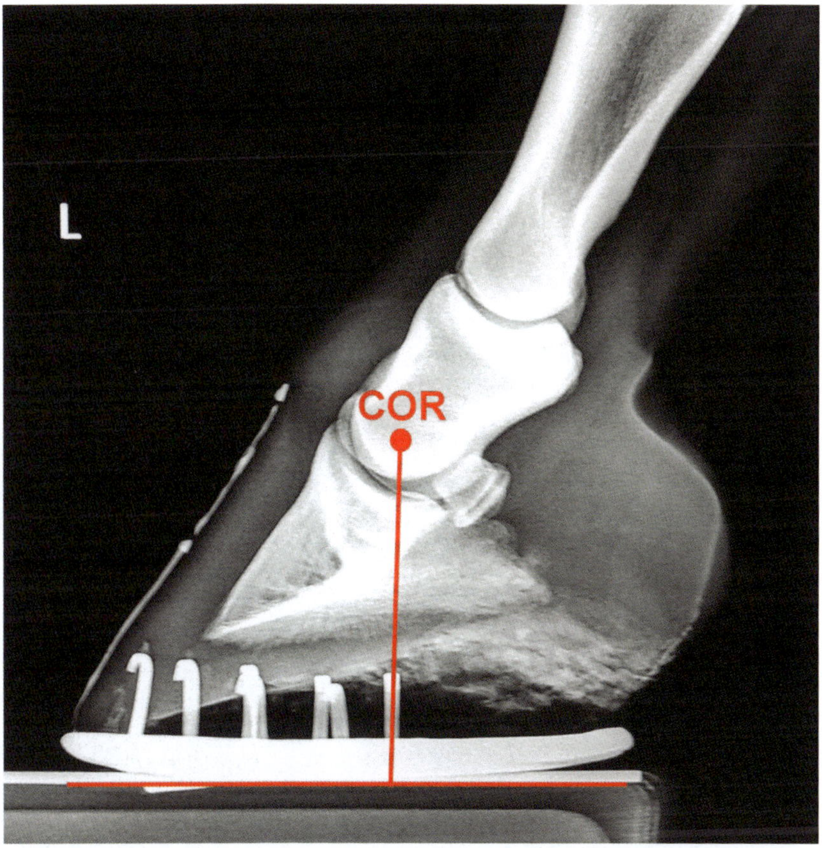

Fig. 6.1: COR is located in the center of the condyle of P2 and is presented as a perpendicular line to the ground.

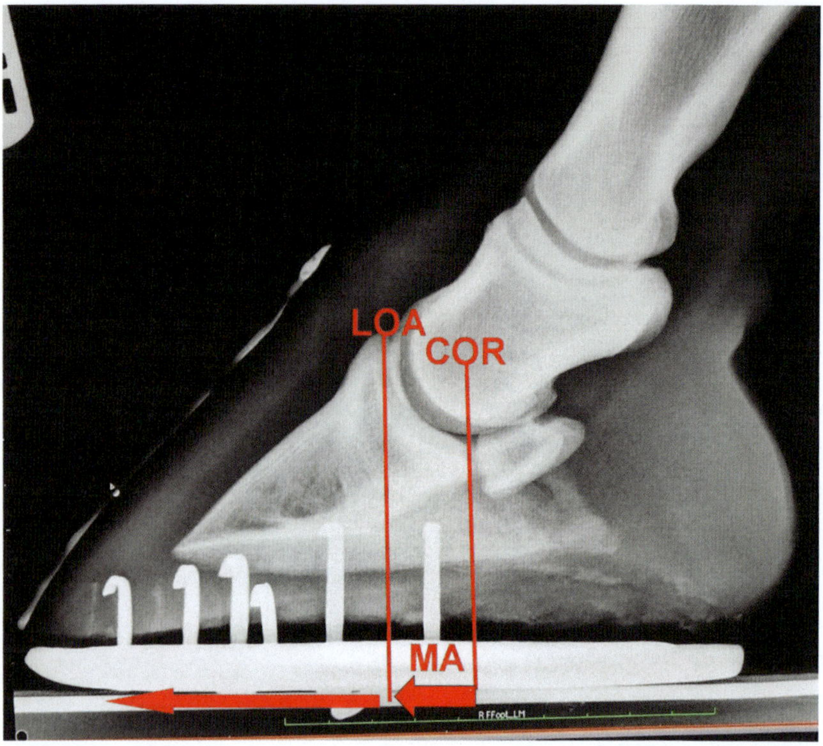

Fig. 6.2: The mechanics of the foot and the distance between the LOA of the extensor and deep flexor tendons to COR at the ground surface of the foot or shoe creates a moment arm (MA) about the coffin joint.

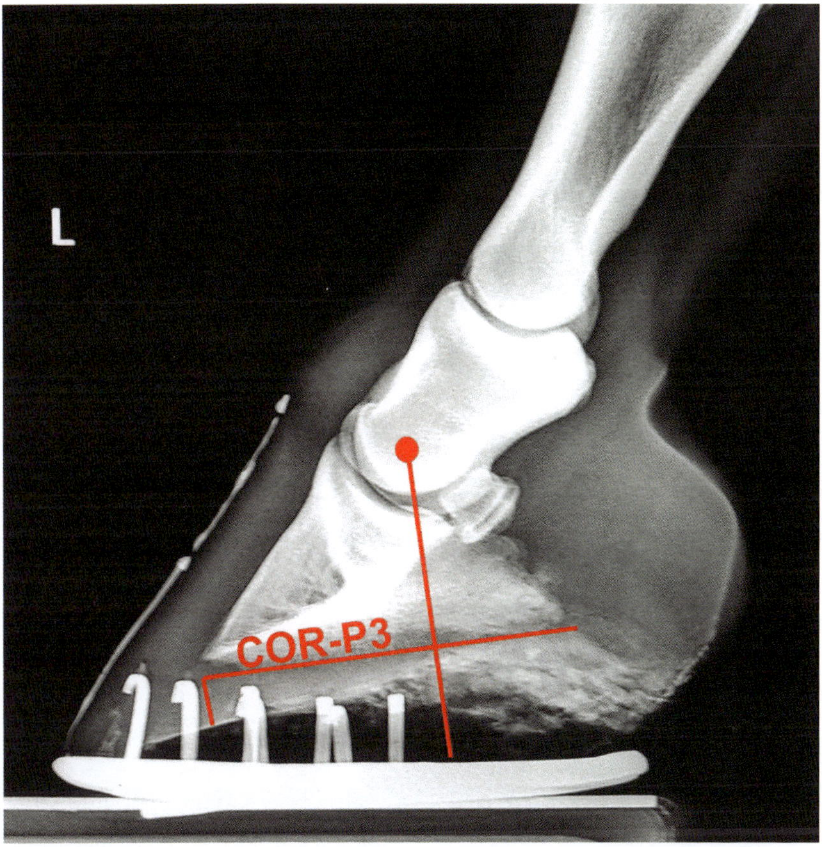

Fig. 6.3: A line drawn from COR at 90° through the ventral plane of P3 to the ground surface of the foot is the COR-P3 vector. The distance between the intersection of the COR-P3 line and palmar P3 to the apex of P3 is the COR-P3 distance.

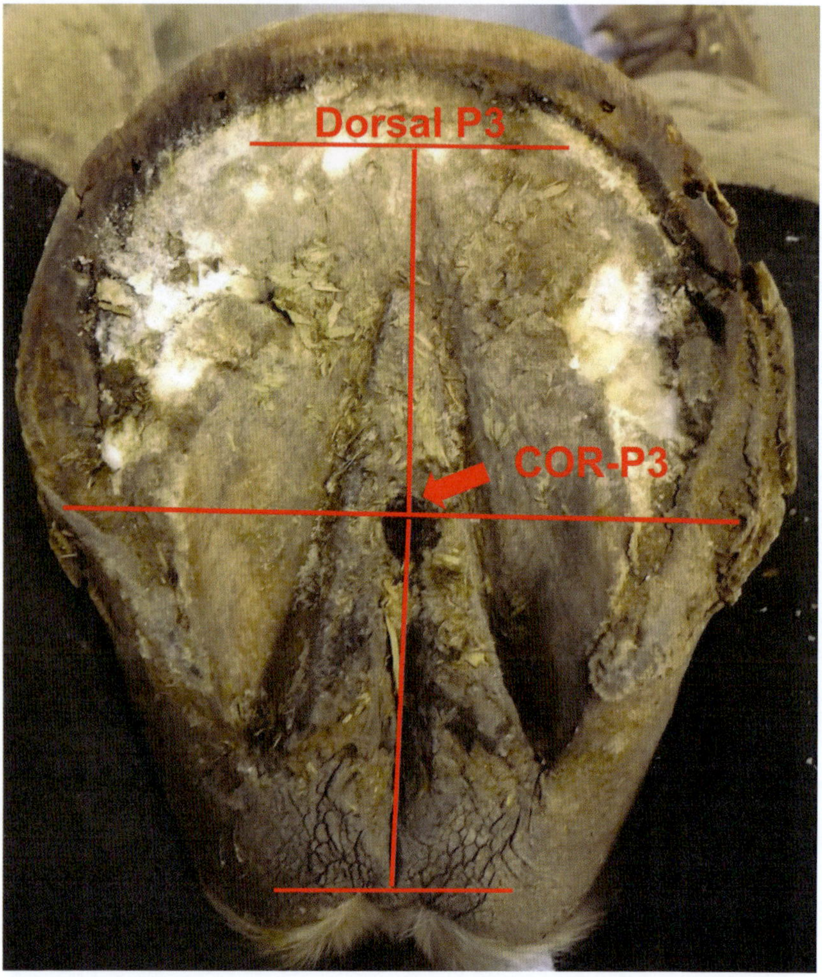

Fig. 6.4: To find the COR-P3 vector: measure dorsally from the insertion of the central frog sulcus at the heel bulb using the COR-P3 distance from the radiograph and mark this point. The COR-P3 vector is a point just dorsal to the central frog sulcus in line with the first bend of the bars. The COR-P3 distance from this point dorsally is the estimated dorsal limit of P3.

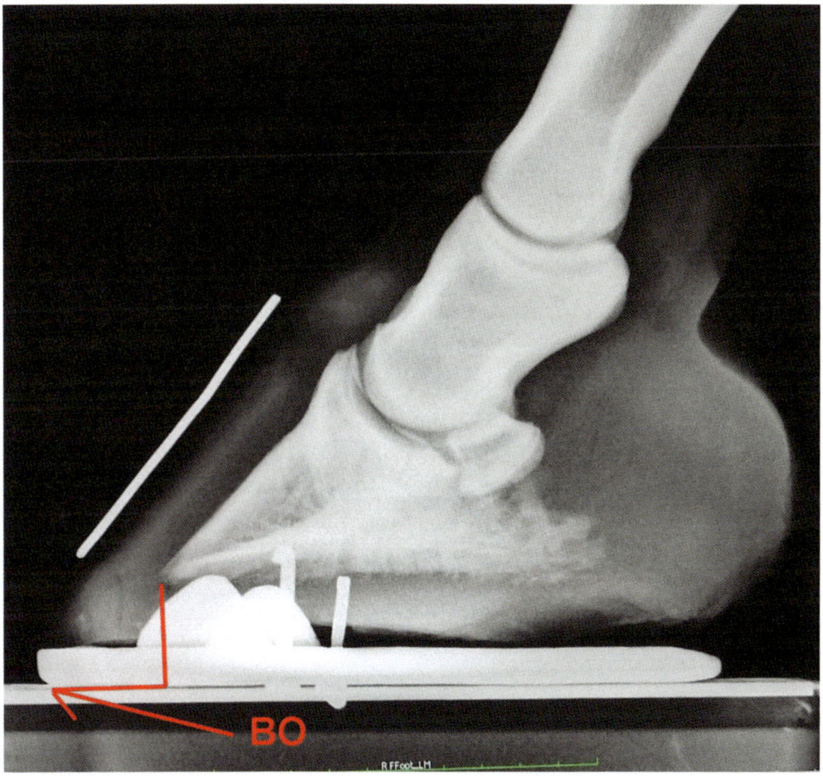

Fig. 6.5: BO is the distance from the most dorsal location of the foot or shoe that contacts the ground to a line from the tip of P3 that is perpendicular to the ventral aspect of P3.

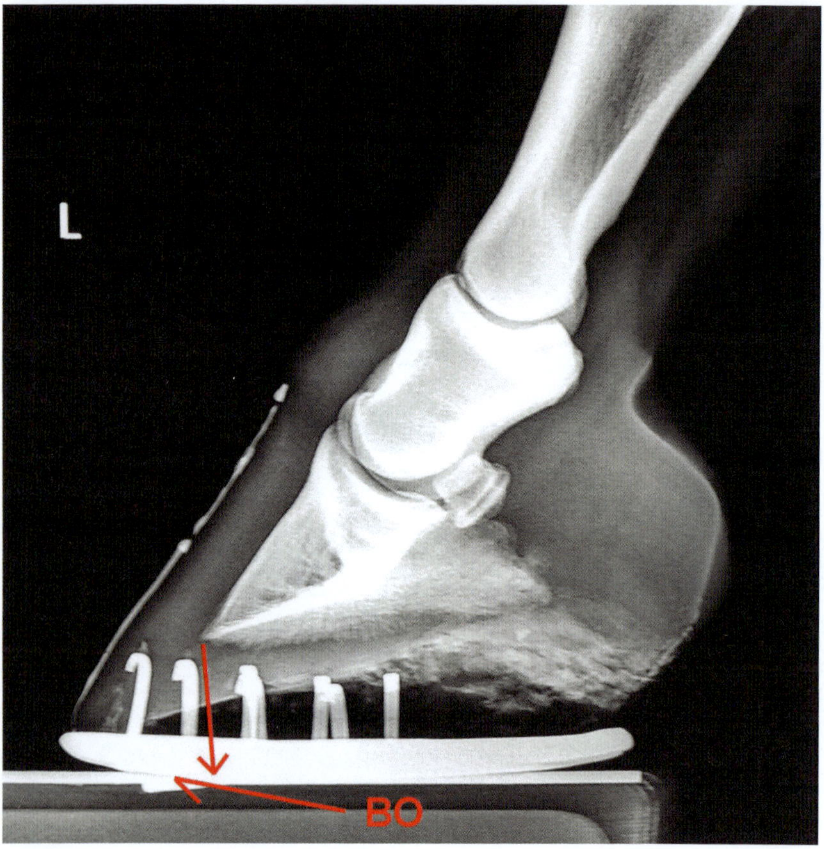

Fig.6.6: Reading the BO on a shoe with a radius can be somewhat ambiguous but provides information as to cranial rotation of the foot onto the toe from the deep flexor tendon tension.

Glossary

Condyle: A rounded protuberance at the end of some bones, forming an articulation with another bone.

Caudal Rotation: As the foot is weighted the vector of COR moves toward the heel creating a lower palmar angle.

Cranial Rotation: As the foot is weighted the vector of COR moves toward the toe creating a higher palmar angle.

Line of Action of the Tendons: The directional relationship of a line drawn from the insertion of the extensor tendon through the insertion of the deep flexor tendon to the ground surface of the foot or shoe.

References

Colles C. Interpreting Radiographs 1. The Foot. Equine Vet J 1983; 15: 297-303.

O'Grady S.E., Poupard D.A., Proper physiologic horseshoeing. Vet Clin Equine 19; 2003: 333-351.

Page B.T., Hagen T.L. Break-over of the hoof and its effect on structures and forces within the foot. Journal of Equine Vet Science 2002; 22: 258-264.

Parks A.H., Aspects of Functional Anatomy of the Distal Limb. In Proceedings: 58th Annual Convention of the American Association of Equine Practitioners 2012: 132-137.

Redden R.F. Hoof capsule distortion: Understanding the mechanisms as a basis for rational management. Vet Clin North Am Pract 19; 2003: 443-462.

Wilson A., Weller R. The Biomechanics of the Equine Limb and Its Effect on Lameness. In: Ross M.W., Dyson S.J., ed. Diagnosis and Management of Lameness Saunders Elsevier, St. Louis Missouri; 2011: 270-281.

7

DORSAL WALL LENGTH, TENDON SURFACE ANGLE AND THE P1-P3 AXIS

Dorsal Wall Length (DWL)

The dorsal wall length (DWL) is the length of the dorsal wall from the top of the coronary band to the ground or shoe surface (Fig. 7.1). This measurement can be transferred to the foot to use as a trimming parameter. Theoretically the DWL should extend distal to the apex of P3 a distance that is 25% of the palmar cortex (PC) of P3, which is the same as a healthy hoof-lamella (HL) zone and sole depth (SD) measurement; applying a radius trim does not negate this distance (Fig. 7.2).

Tendon Surface Angle (TSA)

The tendon surface angle (TSA) is measured as a line along the distal half of the navicular bone to the ground surface (fig. 7.1). This angle can be quite low at 10-15°, or extremely high at 35-40° (Redden 2012). Almost all pathology occurs along the distal half of the navicular bone and its support ligaments. Raising the TSA can reduce friction from the deep flexor tendon (DFT) as it glides over the navicular bone. Pressure on the navicular bone is a function of the tension in the DFT and the angle at which it bends around the bone (Parks 2012). A 1° increase in PA decreases the pressure in the navicular bone by 6% (Eliasher et al 2004).

Monitoring the TSA, along with other balance parameters of the foot, both pre-and post-shoeing, may help in managing navicular pain.

P1 – P3 Axis

The P1–P3 axis is the angle of dorsal P1 as it bisects dorsal P3. Because of the irregular shape of P1, this can be measured by a line that bisects P1 and crosses a line that is parallel to the dorsal face of P3 (Fig. 7.1). To use this measurement as a basis, the cannon bone should be 90° to the ground. Position the horse carefully on the x-ray blocks or the measurement can be distorted. The P1-P3 axis is a valuable measurement in determining proper PA. A 2° PA adjustment will move the P1 -P3 axis 5°.

How the horse postures himself on the blocks can offer information about pain in the feet, noting an extreme vertical or horizontal P1. A normal balanced foot has somewhat of a negative P1-P3 axis of 5-10° as viewed with the foot in the static weight-bearing position, which flexes into a zero-degree axis as the foot is loaded in the weight-bearing position (Healey 2016). The P1-P3 axis is important as it interprets tension of the DFT and the vector of P1 to weight-bearing on the foot through center of rotation (COR) (O'Grady et al 2003). The P1-P3 axis can be measured on the horse as the hoof-pastern axis (HPA) by using a goniometer (Fig. 7.3).

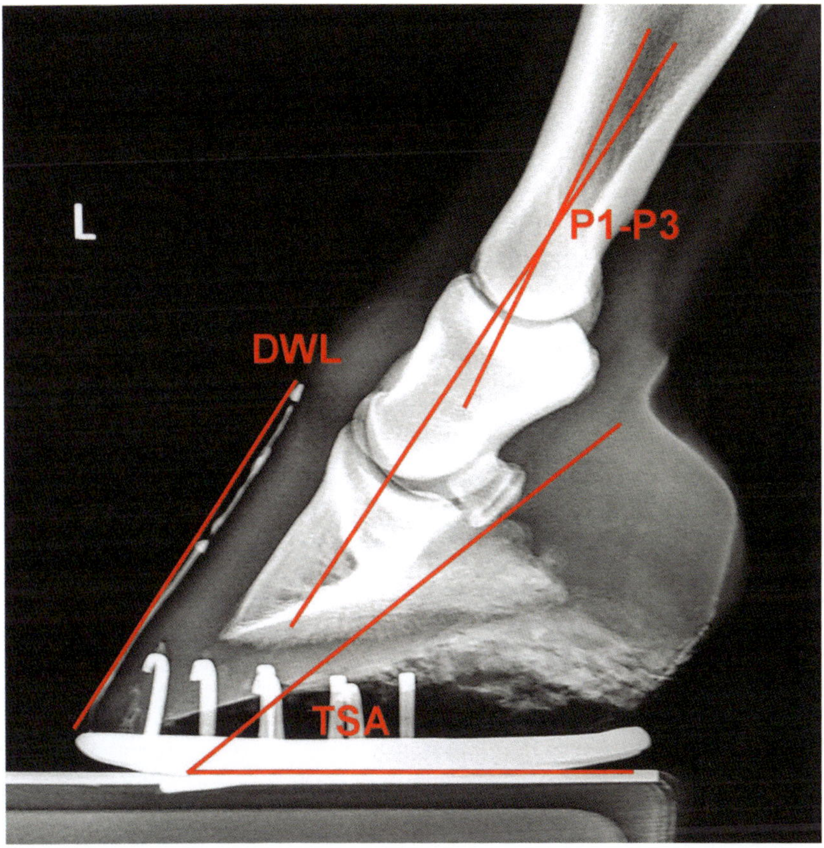

Fig. 7.1: The Dorsal Wall Length (DWL), P1–P3 Axis and the Tendon Surface Angle (TSA) of the navicular bone.

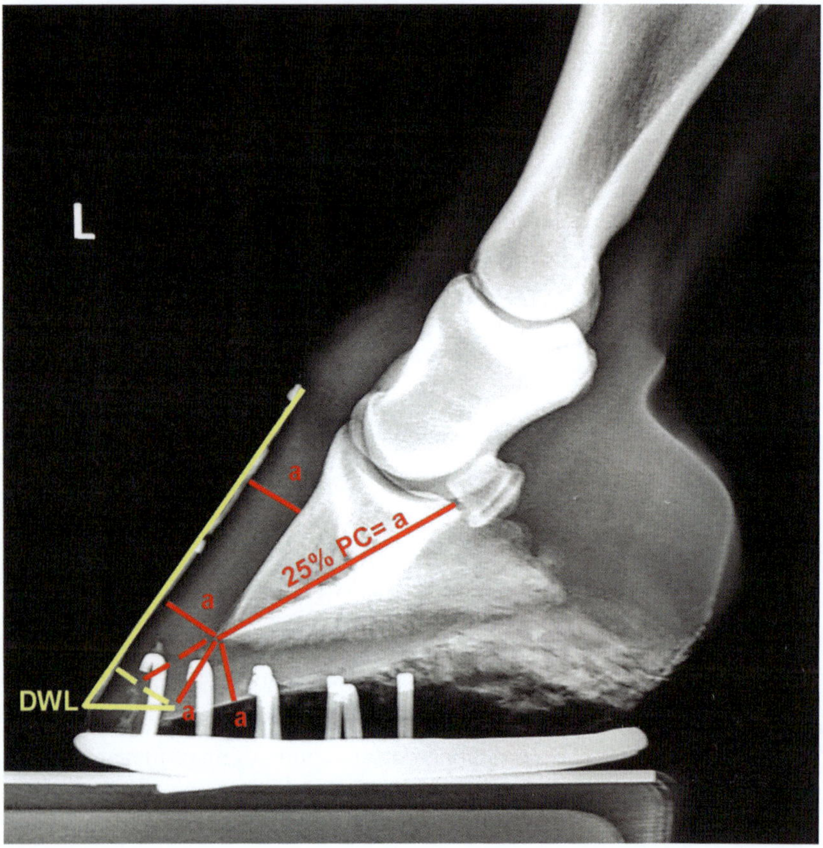

Fig. 7.2: The DWL (yellow line) should extend distodorsal to the apex of P3 a distance equal to 25% of the PC measurement. Applying a radius does not negate this distance (dashed yellow line).

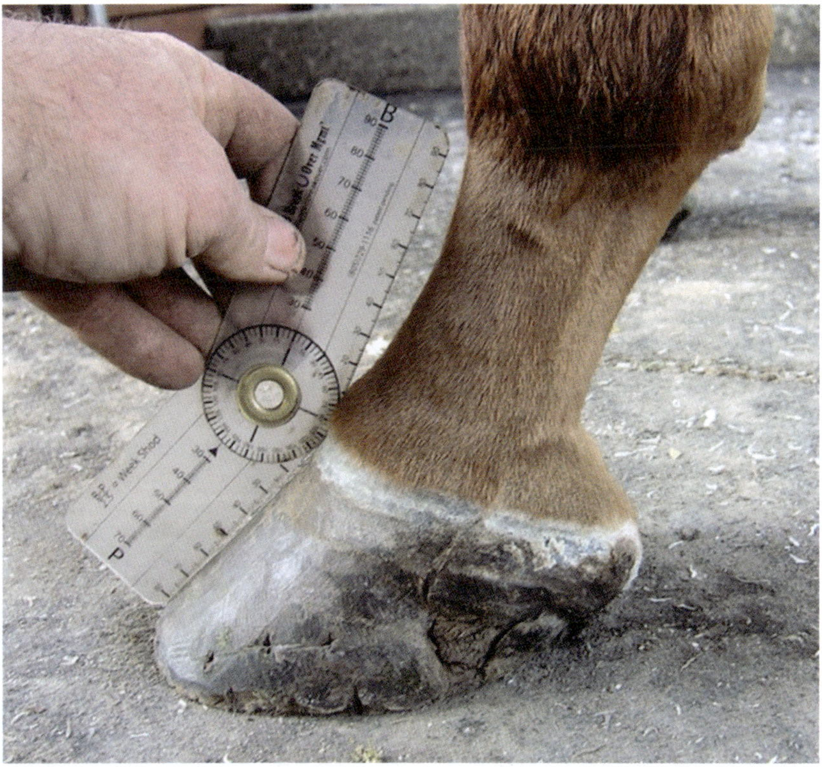

Fig. 7.3: Using a goniometer for hoof-pastern axis (HPA) alignment. Optimum alignment is when the pastern extends to a 0° axis when weight bearing as the opposite foot is non-weight bearing. The HPA on this foot is in the static (non-weight bearing) position and measures -25°. The pastern has a 10° extension in the weight bearing position which would leave a deficit of 15°. A 2° palmar angle (PA) increase will extend the pastern 5°. Therefore, this foot would need a 6° PA increase to balance the HPA.

References

Eliasher E., McGuigan M.P., Wilson A.M. Relationship of foot conformation and force applied to the navicular bone of sound horses at the trot. Equine Vet J. 2004; 36: 431-435.

Healey P. Measurement of the Hoof-Pastern Axis for Foot Management. In: American Farriers journal, Lessiter Media, Brookfield, Wisconsin, December 2016; 50-55.

O'Grady S.E., Poupard D.E. Proper physiological horseshoeing. Vet Clin North Am Pract 2003; 19: 333-334.

Parks A.H. Aspects of functional anatomy of the distal limb. In Proceedings: 58th Annual Convention of the American Association of Equine Practitioners 2012: 132-137.

Redden R.F. Evaluating the Navicular Bone. In: Dr. Redden's Advanced Equine Notes, Redden R.F., Versailles, KY. 2012: 158-167.

8

THE DORSOPALMAR RADIOGRAPH

The dorsopalmar, zero-degree (DP) radiograph is primarily used for evaluating mediolateral balance of the foot. Positioning of the foot is critical for accurate evaluation. Not having the horse positioned squarely, on level ground, will create misalignment of the joints (Little 2007) (Fig. 8.1).

Asymmetry in medial and lateral wall length and angle are common radiographic findings in the DP image. If the proximal interphalangeal and distal interphalangeal joint spaces are congruent, then this asymmetry is not significant (Parks 2011) (fig. 8.2).

Things to Notice

- How straight is the radiograph? Is the extensor process of P3 lined - up with the center of P2?
- Are the joint spaces even?
- If visible, proximal P1 is usually asymmetric
- The higher the arches of the collateral frog sulci, the higher the PA
- P3 is not symmetrical, the medial side is usually a little steeper
- Is the distance from the medial and lateral solar margins equal?
- If the DP balance is really out, it should coincide with a mid-sole lateral projection that shows both wings of P3 (*see* Chapter 1).

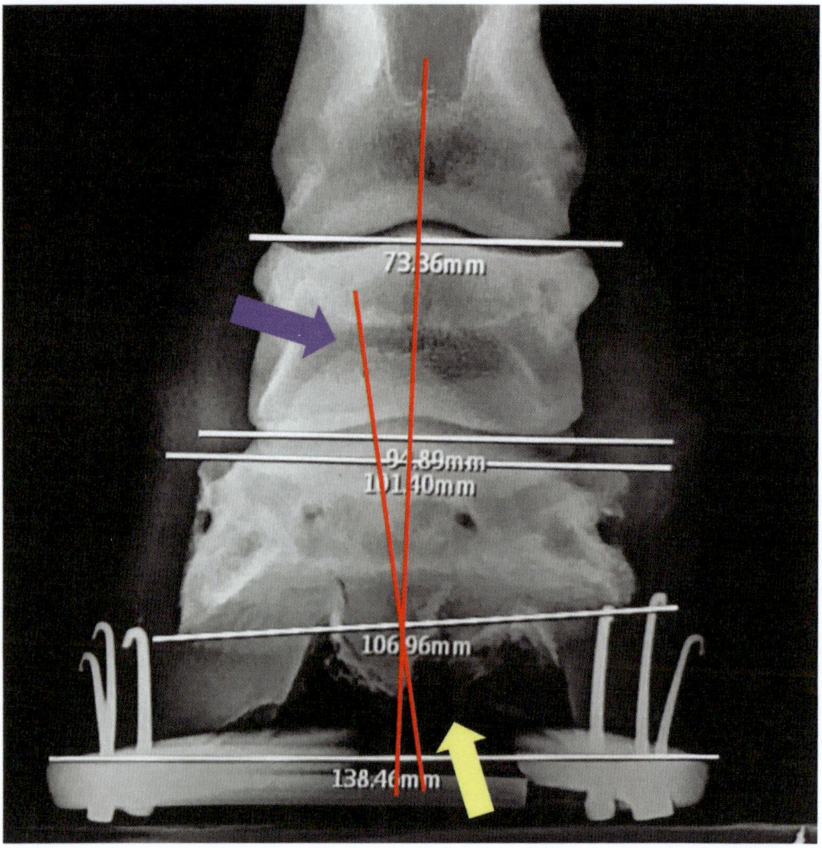

Fig. 8.1: Not positioning the foot properly on the radiograph block or malalignment of the x-ray beam will give a false interpretation of joint and P3 alignment. A correct DP radiograph should show alignment of the extensor process (blue arrow) and central frog sulcus (yellow arrow) with the center line of P1 and P2.

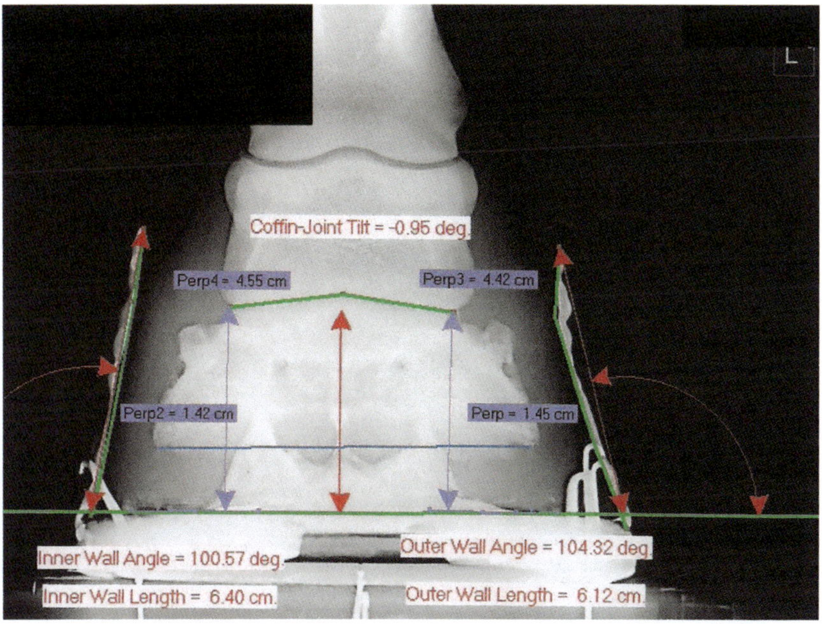

Fig. 8.2: This 'Metron' mark-up of a DP radiograph shows good alignment; note the higher and straighter inter wall length.

Glossary

Distal Interphalangeal Joint Space: The space between the coffin bone (P3) and the middle phalanx (P2).

Proximal Interphalangeal Joint Space: The space between P2 and the proximal phalanx (P1).

References

Little D. The Digital Age and Radiology of The Foot; Diagnostic Evaluation of the Equine Foot. In: Floyd A., Mansmann R.A., ed. Equine Podiatry, St. Louis, Missouri, Saunders Elsevier; 2008: 152-159.

Parks A. The foot and shoeing. In: Ross M.W., Dyson S.J. ed. Diagnosis and Management of Lameness in the Horse. 2nd ed. St. Louis Missouri, Saunders Elsevier; 2011: 282-302.

9

USING CENTER OF ROTATION AS A GUIDE FOR PALMAR ANGLE ADJUSTMENT

Creating balance or equilibrium in the foot requires reducing moment arms (MA). MAs cause leverage which increases load to soft tissues.

As discussed in Chapter 6, the distance between a line drawn from center of rotation (COR) to the ground, to the line of action of the tendons (LOA) to the ground causes a MA about the coffin joint (Wilson et al 2011). This moment combined with other mechanics of the foot has a direct bearing on the amount of load to the heel area. The length of the MA has a relation to the PA. An increase in the PA causes the MA to decrease as the LOA converges toward COR on the solar surface of the foot or shoe. If LOA does not converge into COR at the solar surface of the foot or shoe, this angle difference of LOA to COR can be measured. It will form a line from the proximal aspect of the LOA line at the extensor process, to the distal aspect of the COR line at the ground surface of the shoe or foot (Fig. 9.1).

This author has observed that in a foot with craniocaudal balance, the LOA converges with the line of COR at the ground surface of the foot or shoe (Fig. 9.2). Note that a balanced foot has ample ventral depth, palmarodorsal mechanics and a hoof-pastern axis (HPA) that measures 0° in the weight bearing position. The foot where COR does not converge with the LOA is usually deficient in PA and has a negative hoof-pastern axis (HPA) that does not extend to a 0° axis when weight bearing. This type of foot usually lacks a mechanical radius dorsal to COR on the solar surface of the foot or shoe which prohibits cranial rotation of the foot. The foot where COR over converges the LOA is in a contraction or has a club foot conformation and the HPA overextends several degrees in the weight bearing position.

Using a Formula for PA Adjustment

The formula used for HPA-PA adjustment is that for every 5° that the HPA is negative, add 2° to the PA (Healey 2016). This same formula can be applied to the angle (a) difference between the LOA and COR to give an estimate for PA adjustment.

(a/5)2 = PA Adjustment

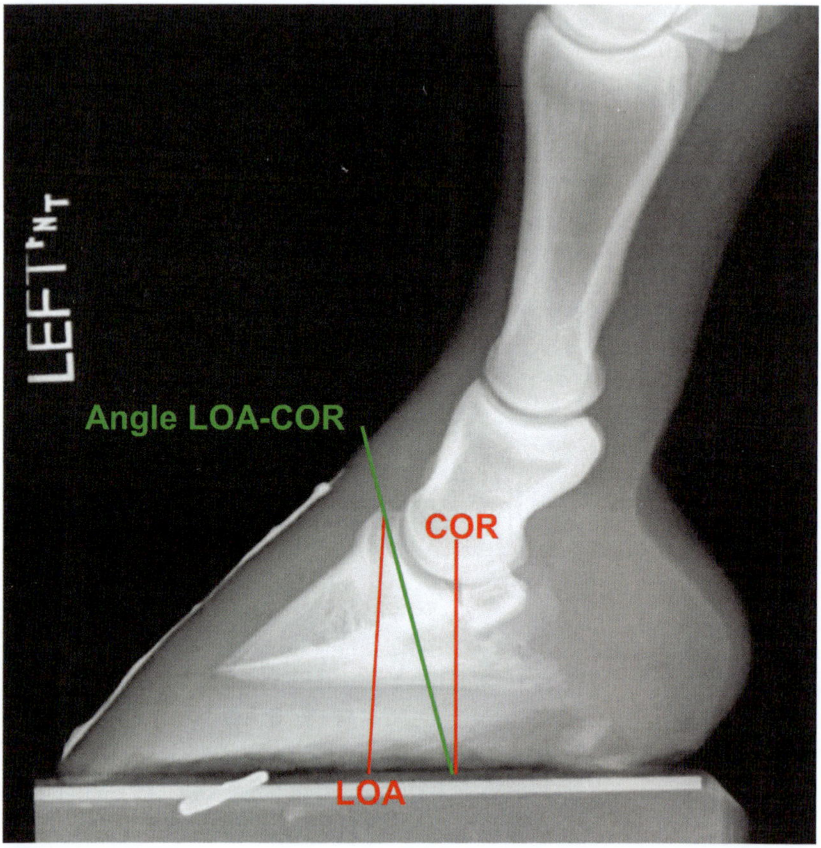

Fig. 9.1: The green line represents the angle difference between LOA and COR.

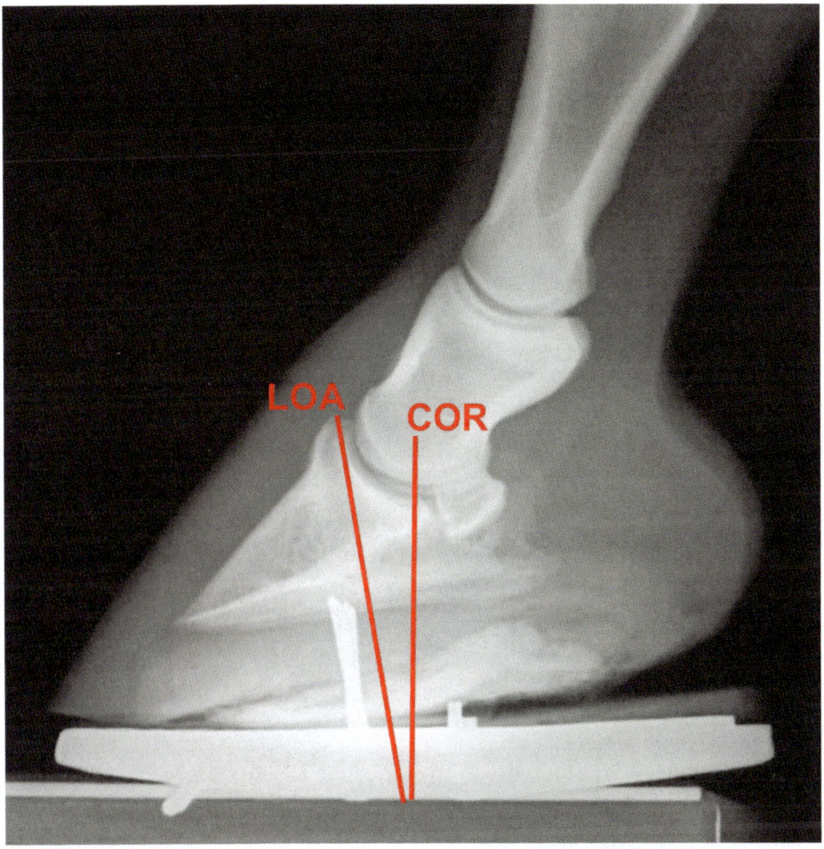

Fig. 9.2: A balanced foot in which LOA converges with COR at the ground surface of the shoe; note the radius in the ground surface of the shoe.

Glossary

Equilibrium: A state in which opposing forces or influences are balanced.
Palmarodorsal: Refers to the heel to toe balance on the solar surface of the foot or shoe.
Radius: A curved line or aspect, not flat.

References

Healey P. Measurement of the Hoof-Pastern Axis for Foot Management. In: American Farriers Journal, Lessiter Media, Brookfield, Wisconsin, December 2016; 50-55.

Wilson A., Weller R. The Biomechanics of the Equine Limb and its Effect on Lameness. In; Ross M.W., Dyson S.J., ed. Diagnosis and Management of Lameness in the Horse, Saunders Elsevier, St. Louis, Missouri; 2011 : 270-281.

10

APPLYING THE MEASUREMENTS: EXAMPLE EVALUATIONS

Example A: Left Front (LF) Foot

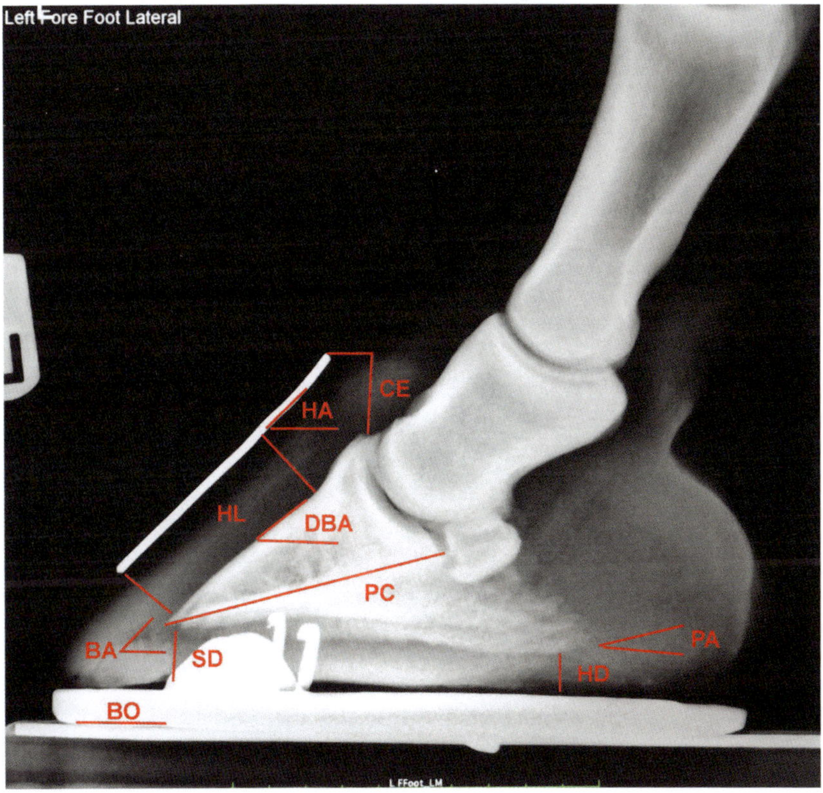

CE: 18 mm	**HD:** 11 mm	**HA:** 46°
HLp: 21 mm	**25% PC:** 20 mm	**DBA:** 40°
HLd: 18 mm	**BO:** 25 mm	**BA:** 43°
SD: 12 mm		**PA:** -3°

The first view impression of the LF foot in example A is a foot that is in a caudal (down and Back) presentation and listed to the medial side (note wings of P3). The coronet-extensor process (CE) distance is considerable at 18 mm. 25% of the palmar Cortex (PC) distance is 20 mm. Using this as a guide, the proximal hoof-lamella (HL) zone is 1 mm thicker but 3 mm wider than the distal HL which is 18 mm. The proximal dermal lamina appears wider which can be verified by the dorsal bone angle (DBA) which is -6° to the proximal hoof angle (HA). Sole depth (SD) is thin: depth at the apex of P3 is 12 mm and at the wing of P3, 11 mm. The palmar angle (PA) is -3° and the break-over distance (BO) is 25 mm dorsal to P3 (leverage/moment arms). Note the bone angle (BA) at 43°, optimum would be 50°.

Example A: Right Front (RF) Foot

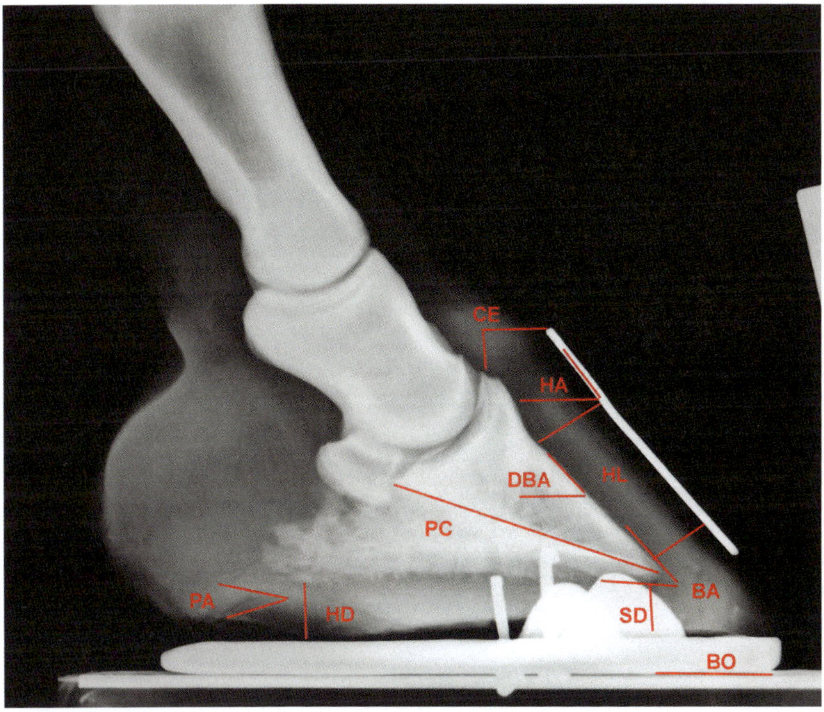

CE: 11 mm	**HD:** 17 mm	**HA:** 52°
HLp: 20 mm	**25% PC:** 20 mm	**DBA:** 47°
HLd: 17 mm	**BO:** 30 mm	**BA:** 47°
SD: 14 mm		**PA:** 0°

Although not as noticeable as the LF, this RF foot in Example A is caudal in the capsule as noted by the DBA-HA difference of 5° and the 3 mm difference between the proximal and distal HL zones. The 0° PA is 3° higher than the LF because of the lift of the deep flexor tendon (DFT), notice the difference in CE measurements of 7 mm. Sole depth is thin at the apex of P3 by 6 mm and heel depth (HD) at the wing is 17 mm which lacks 3 mm plus what would be needed for a correct PA as determined by the alignment of the hoof-pastern axis (HPA) and the degree of moment arm between the line of action (LOA) of the deep and extensor tendons to center of rotation (COR).

Example B: Left Front (LF) Foot

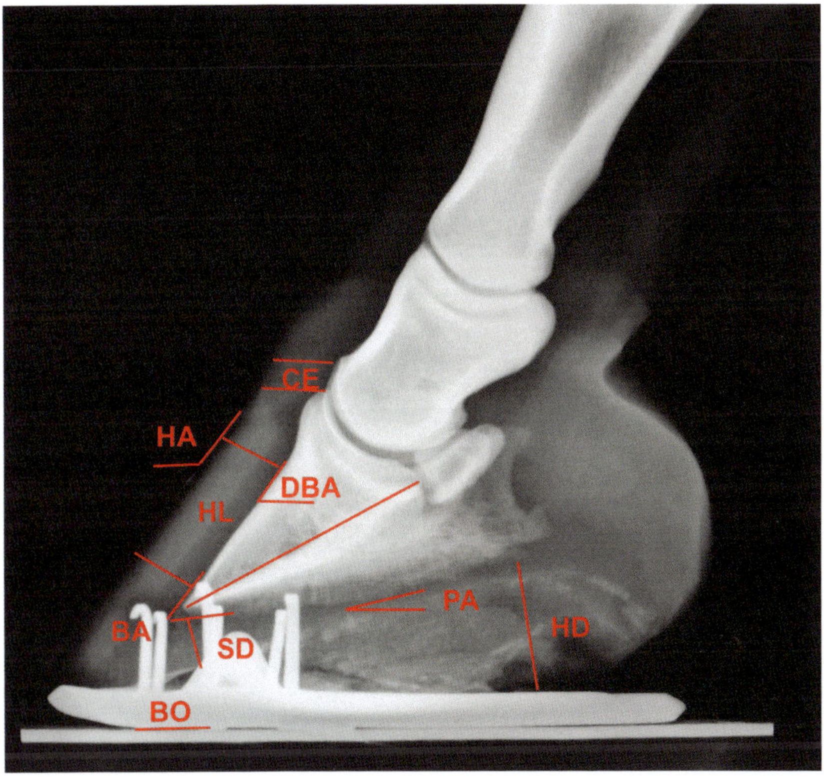

CE: 9 mm	**HD:** 37 mm	**HA:** 55°
HLp: 19 mm	**25% PC:** 19 mm	**DBA:** 55°
HLd: 20 mm	**BO:** 20 mm	**BA:** 45°
SD: 15 mm		**PA:** 10°

The opposite of the feet in Example A, this foot is in a cranial rotation, a product of more suspension from the DFT. Notice the difference in the CE and HL zones between feet. There is a slight distortion of the distal HL zone of this foot due to toe leverage. The toe leverage has been reduced to 20 mm by the use of a rocker toe but because of the high PA of 10°, this places the line of action of the apex of P3 in a more palmar position. The high PA is a necessity as the BA is only 45°. The BA (45°) + PA (10°) = HA (55°), which is physiologically correct for this limb. Note the HPA at about -10°. The higher PA places more weight toward the apex of P3. Combined with the leverage arm of the BO distance, this can have an effect on the solar corium and subsequent sole depth; SD is at 15 mm, HD is at 37 mm.

Example B: Right Front (RF) Foot

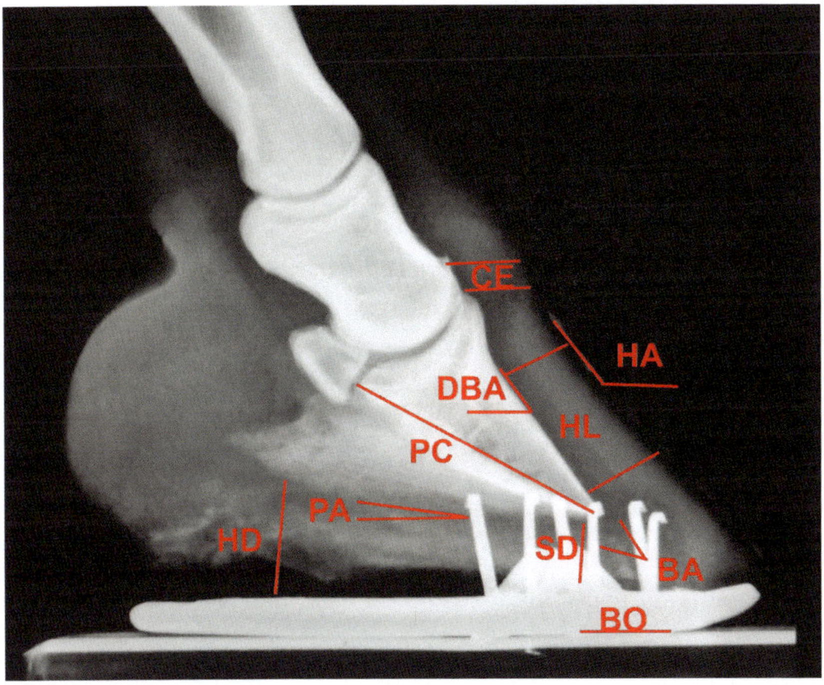

CE: 7 mm	**HD:** 28 mm	**HA:** 53°
HLp: 19 mm	**25% PC:** 19 mm	**DBA:** 55°
HLd: 27 mm	**BO:** 23 mm	**BA:** 48°
SD: 17 mm		**PA:** 7°

Notice the posture of the HPA, which is very straight indicating tension in the deep flexor tendon-muscle unit, although the PA in this foot of 7° is 3° lower than the LF which is at 10°, the BA of the RF is 48° which is 3° higher than the LF at 45°. Because of the toe distortion of 27 mm in the distal HL zone, this presents a HA of 53° which is 2° lower than the LF, note the difference in the DBA to HA angles as P3 is slightly rotated. Using 25% of the PC, SD is 2 mm too thin and the HLd is 8 mm too wide.

11

USING THE INFORMATION TO MANAGE FEET: SHOEING EXAMPLES

These shoeing examples utilize the radiograph information to reestablish equilibrium to the feet. A mechanical application of the shoe package will help rehabilitate the foot during grow-out by minimizing distortion leverages. These are real life clinical examples from horses that presented varying degrees of lameness. Post shoeing follow-ups revealed substantial improvement in mitigating lameness. It must be noted that lame horses do not become sound immediately after shoeing. Improvement in soundness may be seen within a few days of shoeing, depending on the duration of the lameness, injury and inflammation; sometimes it will take longer.

The shoeing methods used by this author are just one way to accomplish the mechanical needs of the foot. There may be several different shoe-pad applications that achieve the same goals. It is important for the veterinarian, farrier and owner to have a conversation regarding shoeing options and expectations. The soft tissue parameters and bone angles provide specific direction for therapy goals.

(Case Study 1)

Case 1: 25% PC: 16 mm; CE: 6 mm; HLp: 17 mm; HLd: 16 mm, SD: 19 mm; HD: 20 mm; BA: 43°; HA: 42°; DBA: 39°; PA: -4°; BO: 40 mm

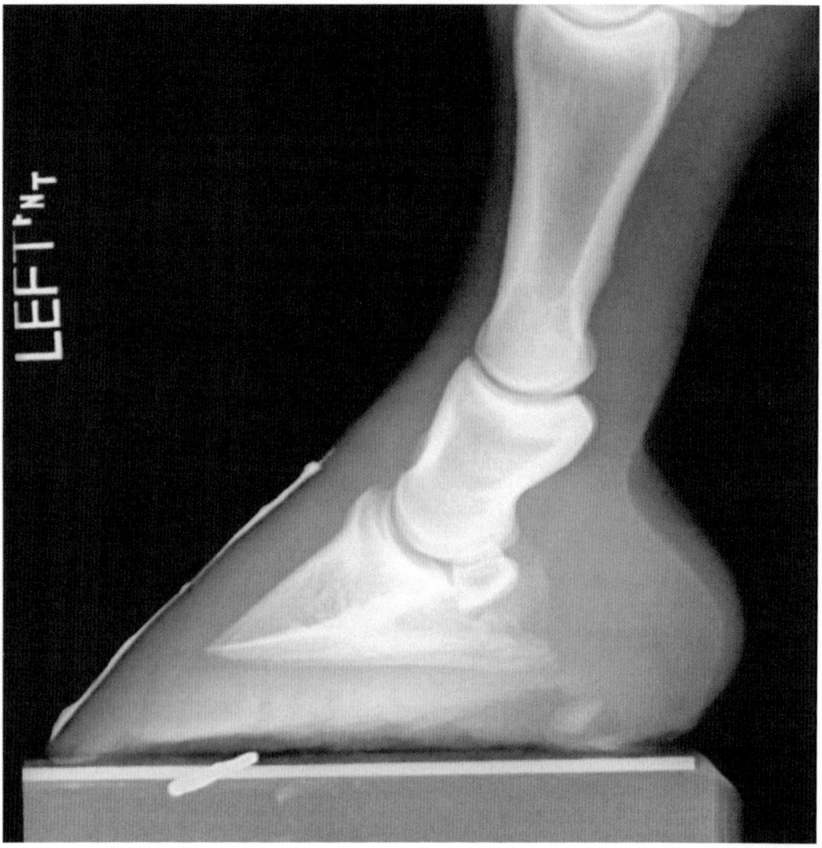

This LF foot (fig. 11.1) is caudally rotated and has a palmar angle (PA) of -4°. The dorsal bone angle (DBA) is 3° negative to the hoof angle (HA) and the proximal hoof-lamella zone (HL) is wider than the distal HL. The angle difference between the line of action of the tendons (LOA) and center of rotation (COR) is 20° thus (20/5)2 = 8° as an estimated PA increase. There is surplus hoof between the apex of P3 and the apex of the hoof that can be rockered back to COR without losing optimum sole depth (SD) of 16 mm. Removing this toe wedge will immediately increase the PA several degrees. Separations caused by the combined result of the minimum heel depth (HD) and over-loading of the heels are seen as an air gap in the quarters.

Goals:

- Reduce the moment arm of the toe; place the BO radius palmar to the tip of P3 + - 30 mm to reduce leverage to the dorsal laminar complex
- Raise PA + - 8° to minimize leverage at the coffin joint and load at the heels
- Rebuild the heel area to increase HD and to protect the existing heel from abrasion

During shoeing (Fig. 11.2) a second radiograph of the LF was taken to evaluate the mechanical goals. Because the palmar half of the foot was weak, an acrylic was added for depth and to reduce shoe abrasion to the existing heels. This extra depth allows the foot to be rockered into a cranial rotation; new HD measurement 25 mm. During shoeing, the hoof-pastern axis (HPA) is monitored with a goniometer to estimate PA alignment. The negative 3° between the DBA and HA is subtracted from the goniometer measurement for accurate evaluation. A 2° wedge pad was added to the rocker shoe and then the shoe was set with two nails for a radiograph evaluation before final changes and nail-up. The radiograph shows that alignment is close, the LOA is within 3° of COR. Using the formula: (3/5)2 = 1.2°, an additional 1° wedge pad can be added before nail-up of the shoe or because there is still ample sole depth of 19 mm, an additional 3 mm could be rockered off the toe. The bone angle (BA) of 43° plus the new PA of 8° would equal a 51° foot. Notice the ground bearing surface of the shoe and the mechanics which reduce ground reaction forces at the toe and heel (red arrows).

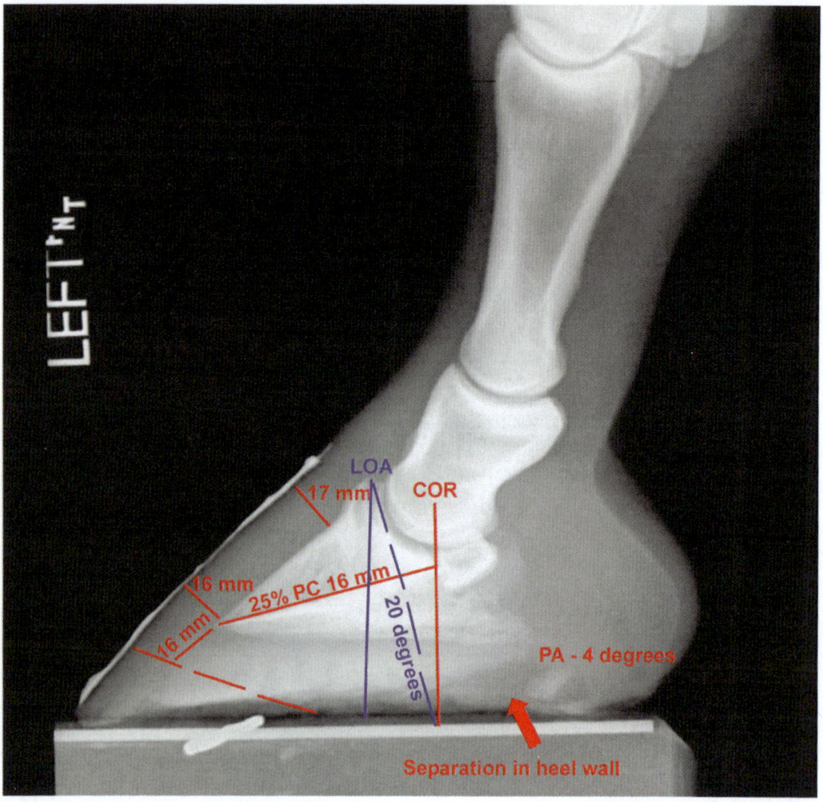

Fig. 11.1: Pre-shoeing radiograph of Case Study 1.

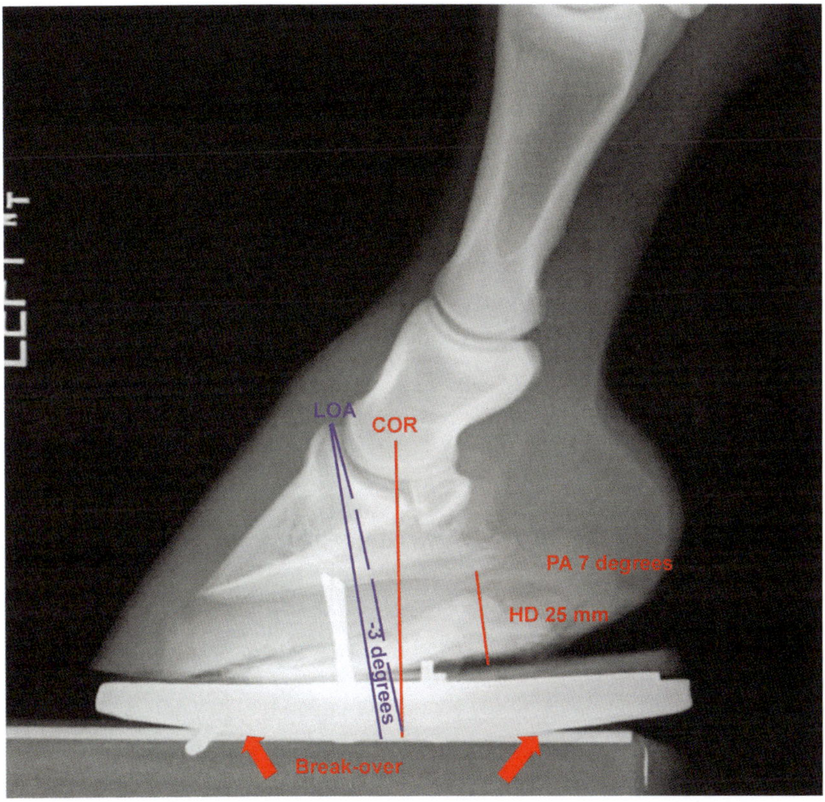

Fig. 11.2: Interim shoeing radiograph of Case Study 1.

(Case Study 2)

Case 2: 25%PC: 17 mm; HLp: + - 20 mm; HLd: + - 17 mm; SD: 19 mm; HD:28 mm; CE + - 7 mm; BA: 45°; HA: 53°; DBA: 49°; PA: 4°; BO: 20 mm

(Note: Because of the absence of a dorsal wall marker, the HL zones and CE distances are somewhat ambiguous and measured as + -)

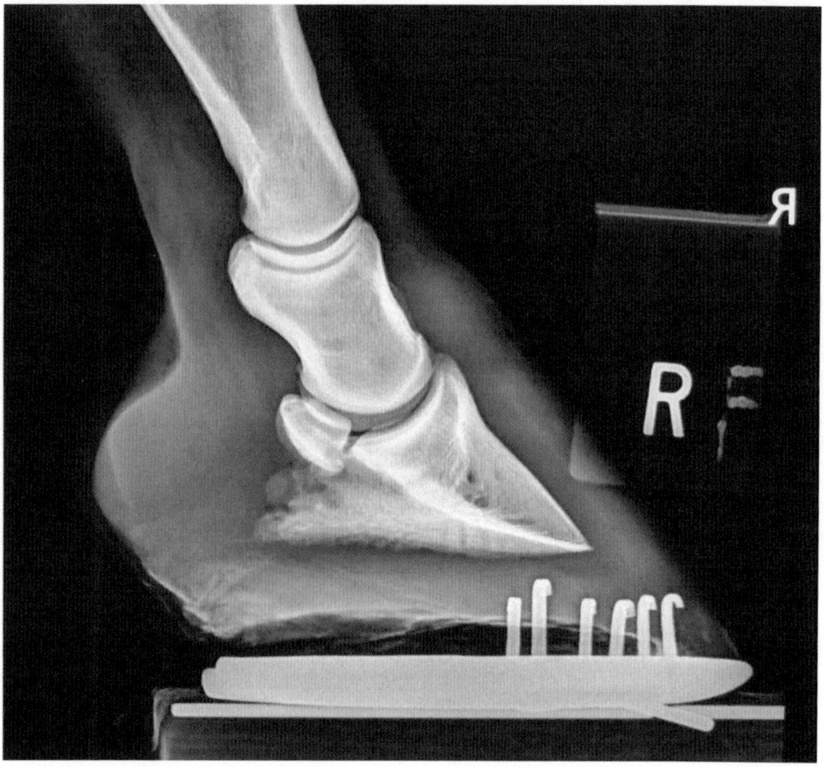

The preshoeing radiograph of this RF foot appears somewhat normal with a PA of 4°. There is ample sole depth at the apex and the wing of P3 (Fig. 11.3). BO is + - 20 mm dorsal to the line of action of the apex of P3 and the hoof capsule and shoe are displaced dorsally, notice how much shoe (green line) is dorsal and palmar to the line of COR. The proximal interphalangeal joint is slightly misaligned (red arrow) and P1 is vertical to P3. The lack of a dorsal wall marker makes accurate HL zones and CE measurements difficult. The angle difference between LOA and COR is 15° which estimates a 6° PA increase (15/5)2 = 6°.

Goals:

- Create mechanical break-over through a roller-motion/ rocker shoe to the widest part of the foot to decrease leverage to the proximal lamina.
- Increase PA 6°+ degrees using the HPA as a guide and taking into consideration the negative DBA-HA angle.
- Realign ground bearing surface of the shoe with COR.

The post shoeing radiograph (Fig. 11.4) has a dorsal wall marker which makes measurement of the HL zones more accurate and adds an extra 2 mm to each zone. This is significant as a healthy HL zone for this foot would be 17 mm and this radiograph measures 22 mm. This extra 5 mm is in the lamina side and when combined with the negative DBA – HA of 4° indicates lack of structural integrity to the proximal lamina. Excess toe was rockered off to COR and the heels were leveled palmar to COR at the height of the untrimmed frog. A goniometer was used to monitor the foot for HPA – PA alignment, taking into account the negative DBA – HA angle. The foot was shod with a roller-motion (rocker) shoe with 3°s of wedge pad, new PA 13°. Note how LOA converges with COR at the ground surface of the shoe. Because of the radius of the shoe being close to COR, this allows the tension of the deep flexor tendon-muscle unit to achieve equilibrium in the foot by rotating the foot onto the toe (dashed blue arrowed line), note the continuum of the radius of the shoe at this point.

BA 45°+ PA 13° = foot of 58° (HA 62°)

Figure 11.5 shows the same foot 8 weeks post shoeing. The PA has reduced to 10° but still retains equilibrium between the LOA and COR because of the added sole depth during grow-off. SD: 31 mm (+12 mm), HD: 36 mm (+8 mm).

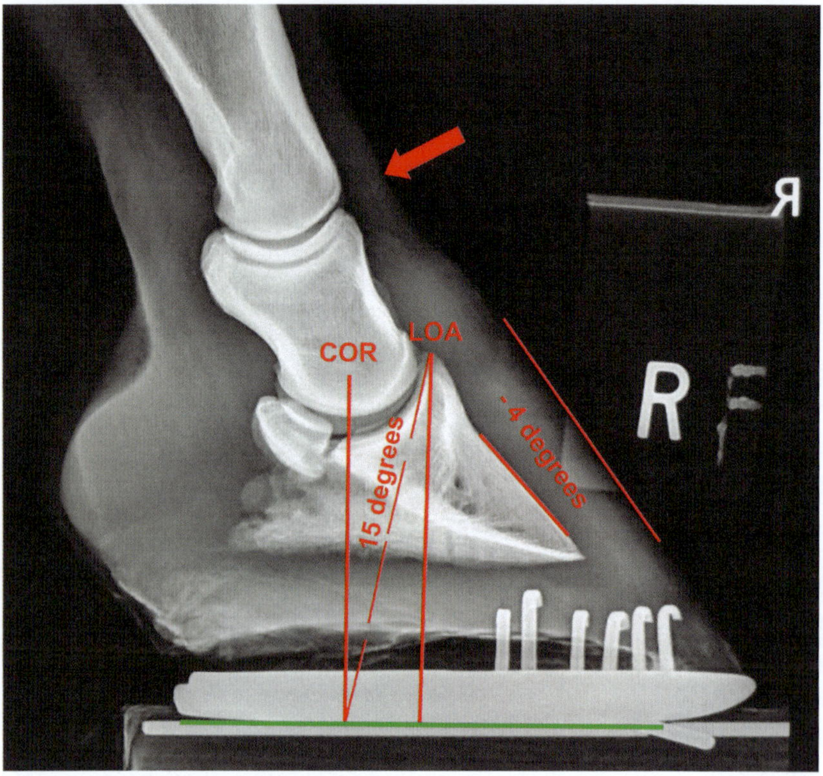

Fig. 11.3: Pre-shoeing radiograph of Case Study 2.

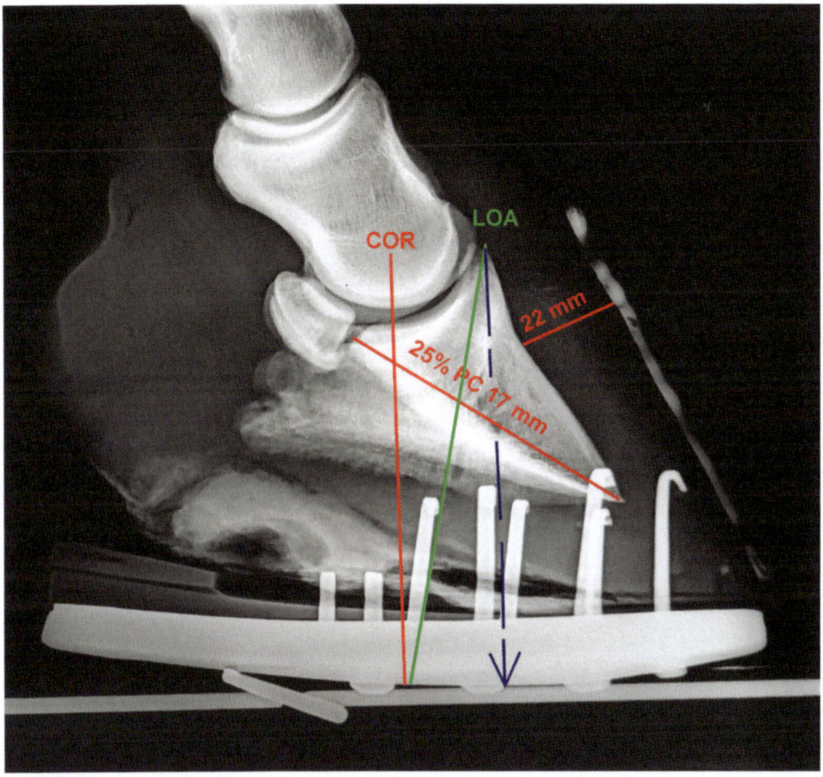

Fig. 11.4: Post shoeing radiograph of Case Study 2.

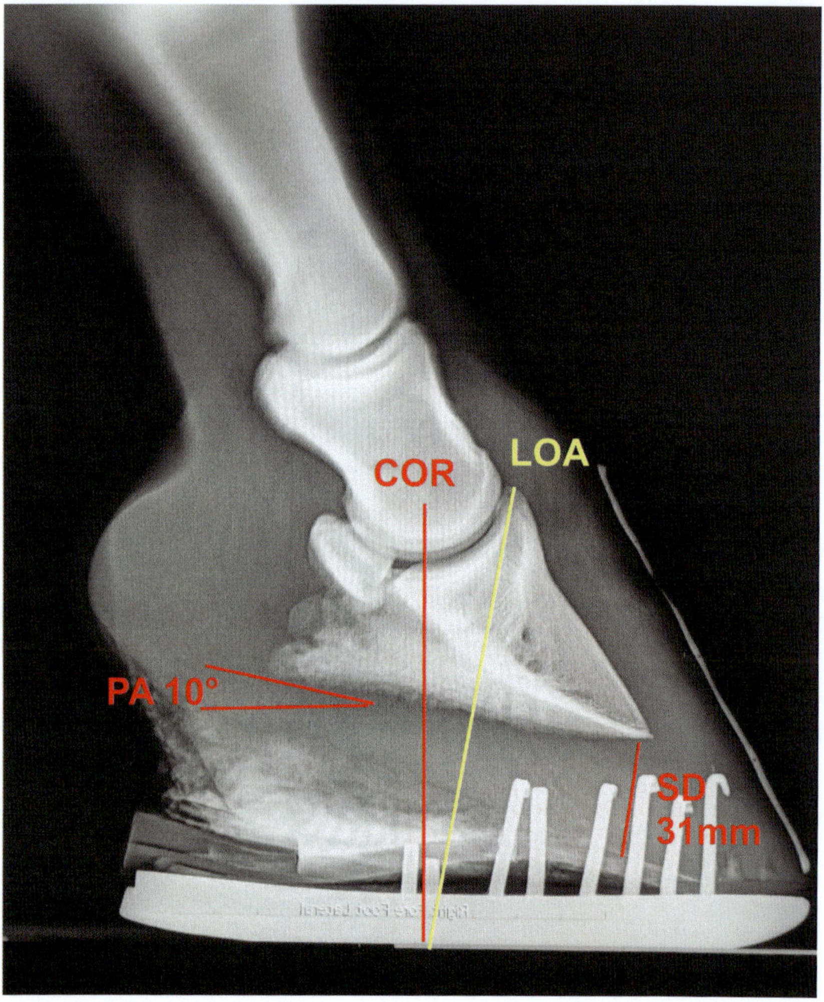

Fig. 11.5: Radiograph of Case Study 2, eight weeks post shoeing. The SD distance is measured to the solar surface of the pad.

(Case 3)

Case 3: 25% PC: 17 mm; HLp: + - 19 mm; HLd: + - 17 mm; SD: 18 mm; HD: 24 mm; CE + - 7 mm; BA: 46°; HA: 53°; DBA: 47°; PA: 1°; BO: 21 mm

(Note: Because of the absence of a dorsal wall marker, the HL zones and CE distances are somewhat ambiguous and measured as + -)

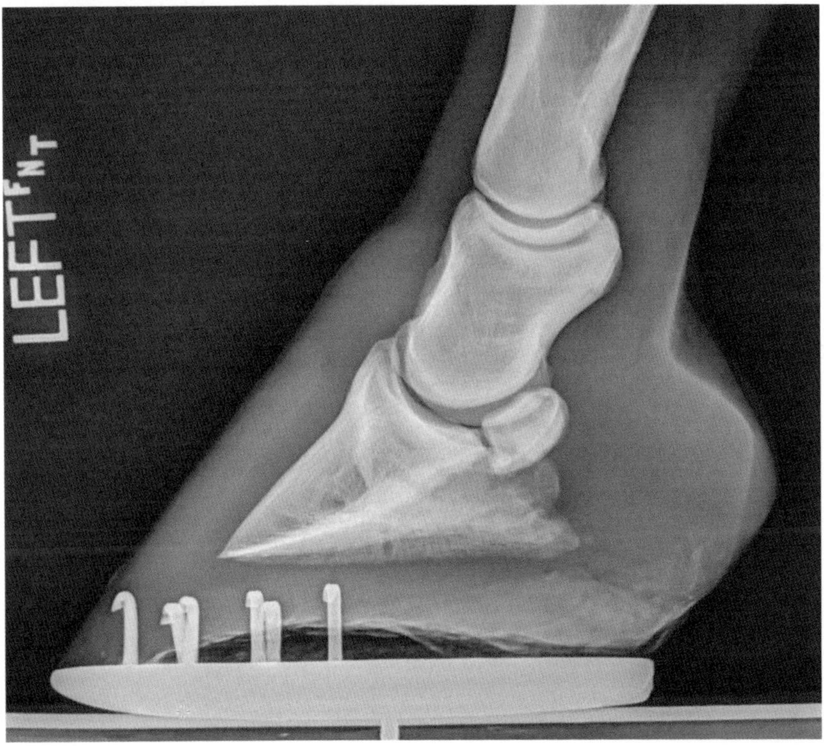

This LF foot (Fig. 11.6) is from the horse in Case 2. Although not always the norm, this foot is very similar to the RF. The PA of 1° compared to the PA of 4° in the RF is due to less lift from the deep flexor tendon (DFT). This is reflected in the DBA-HA which is -6° compared to the RF at- 4°. The BA is 1° greater than the RF; break-over is the same at + - 20 mm. The knuckling of the interphalangeal joint (red circle) may be due to superficial flexor tension.

Like the post shoeing radiograph of the RF, the dorsal wall marker on the LF (Fig. 11.7) gives a more accurate measurement of the HL zones adding 2 mm. The LOA cranially rotates the foot onto its widest part (dashed red line). Note the radius of the ground surface of the shoe is directly underneath the extensor process of P3 (red arrow), relieving ground reaction forces (GRF) to the dorsal complex of the foot. The radius of the shoe also prevents heel crushing from the GRF palmar to the line of COR-P3 (yellow arrow). This foot required the same amount of adjustment as the RF and produced a PA of 12°. Note how the internal angles produce the same angled foot.

RF: BA 45° + PA 13° = 58°

LF: BA 46° + PA 12° = 58°

At 8 weeks post shoeing (Fig.11.8), the PA has reduced to 8° from 12°. SD: 26 mm (+9 mm), HD: 32 mm (+11 mm). LOA-COR maintains equilibrium; note ground mechanics of the shoe (red arrow).

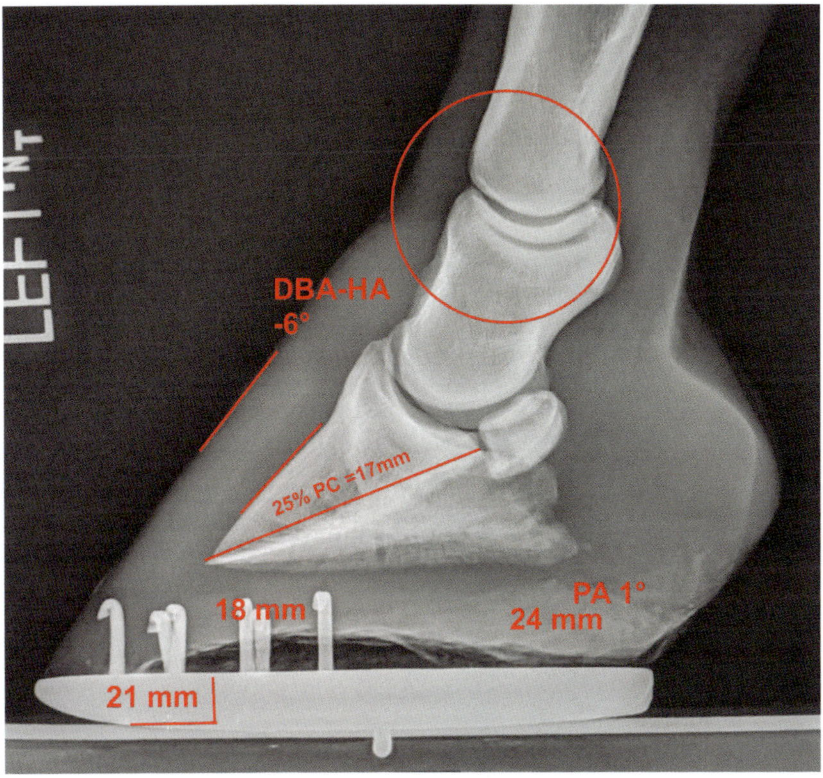

Fig. 11.6: Pre-shoeing radiograph of Case Study 3.

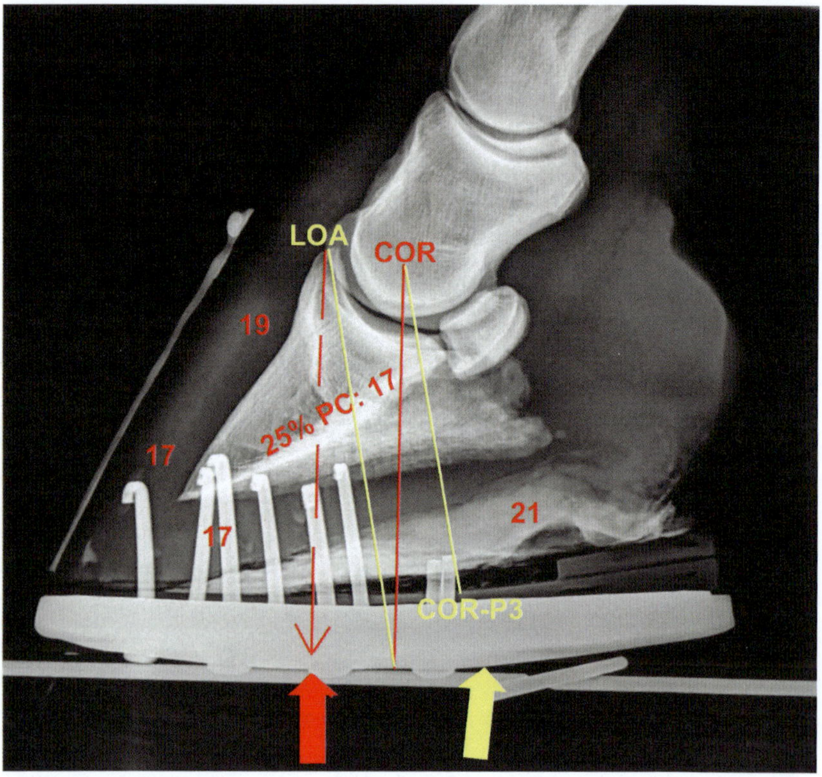

Fig. 11.7: Post shoeing radiograph of Case Study 3.

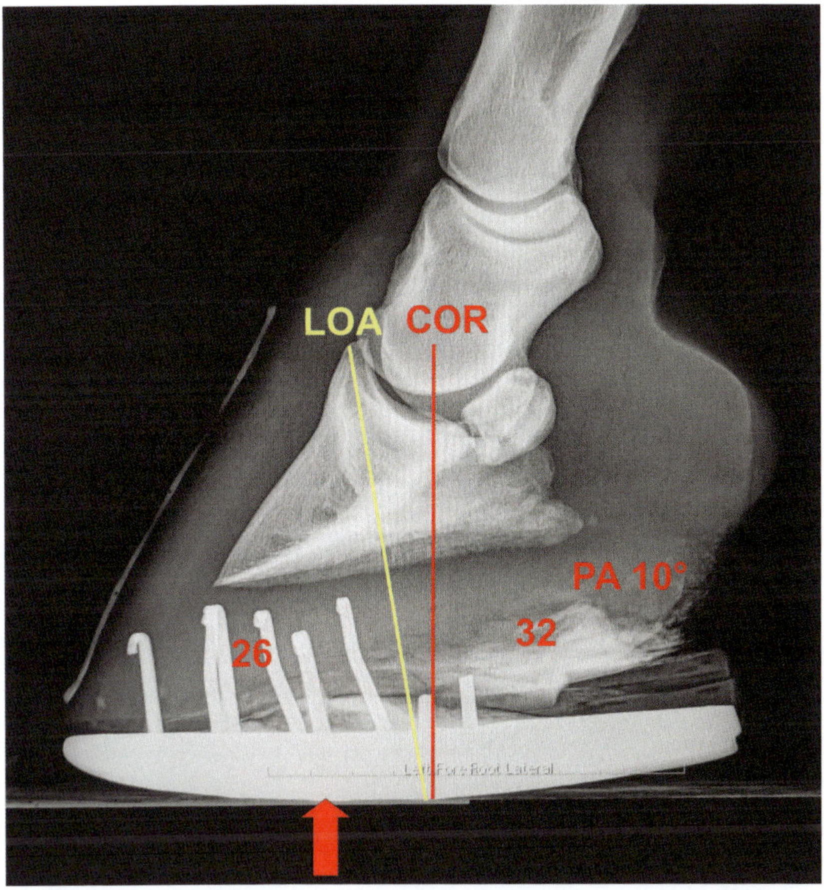

Fig. 11.8: Radiograph of Case Study 3 at eight weeks post shoeing. SD is 26 mm and HD is 32 mm. This foot is still mechanically balanced.

(Case 4)

Case 4: 25% PC: 19 mm; HLp: 20 mm; HLd: 24 mm; SD: 16 mm; HD: 13 mm; CE: 13 mm; BA: 43°; HA: 42°; DBA: 42°; PA: 0°; BO: 27 mm

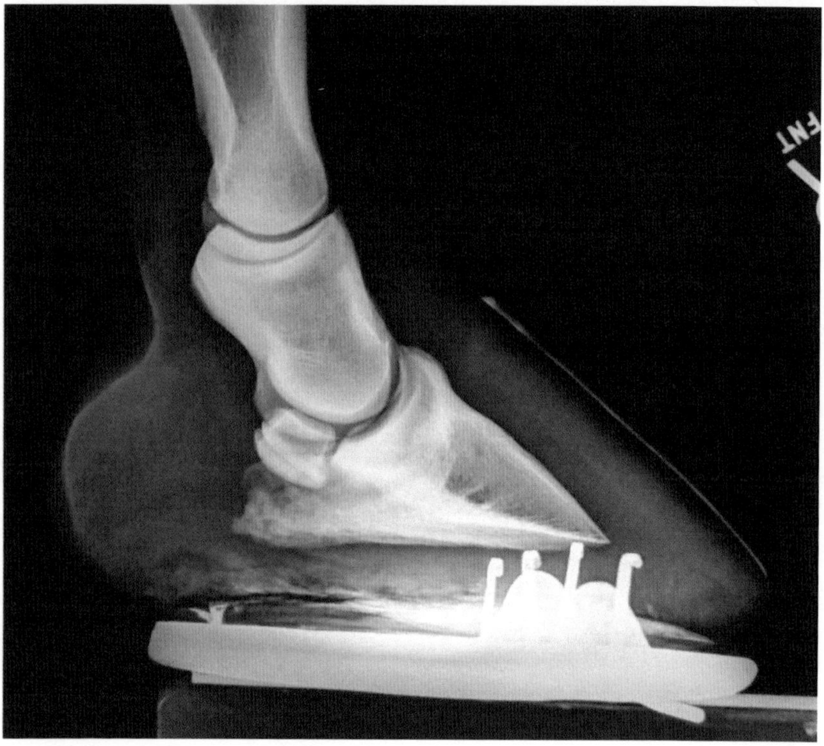

The pre-shoeing radiograph of this RF foot (Fig. 11.9) displays a flat foot with low ventral heel mass. There has been an attempt to correct this foot by shortening the toe and wedging the heel. The PA of 0° is actually -1° to the shoe package. The HD of 13 mm cannot sustain the wedge because of the deterioration of the heel mass which is a product of the leverages dorsal to COR. The LOA-COR is 15° (dashed green line) and BO is 27 mm. Although the shoe is set back, the mechanics of it are flat. The straight position of P1 distributes weight palmar to COR. The white line area is thickened (blue arrow). This correlates to the thickened distal HL zone. Although this foot is in a caudal rotation, it is not caudally displaced; HA-DBA = 0°. The BA of 43° shows that P3 is 1° negative to the shoe package.

Goals:

- Build ventral depth
- Reduce break-over leverage
- Reduce moment arm of LOA-COR
- Increase PA as needed for craniocaudal balance, using HPA as a guide

The post shoeing radiograph (Fig. 11.10) shows the moment arm of LOA-COR is reduced to 0°. Break-over leverage has been reduced to the line of action of the apex of P3, GRF at the heel has been eliminated. Equilibrium in ventral depth has been created using an acrylic build making SD 19 mm and HD 21 mm. Craniocaudal balance was established by monitoring the HPA with a goniometer at the shoeing floor. The new PA of 10° is a combination of a 6° wedge shoe/pad package and a rocker shoe.

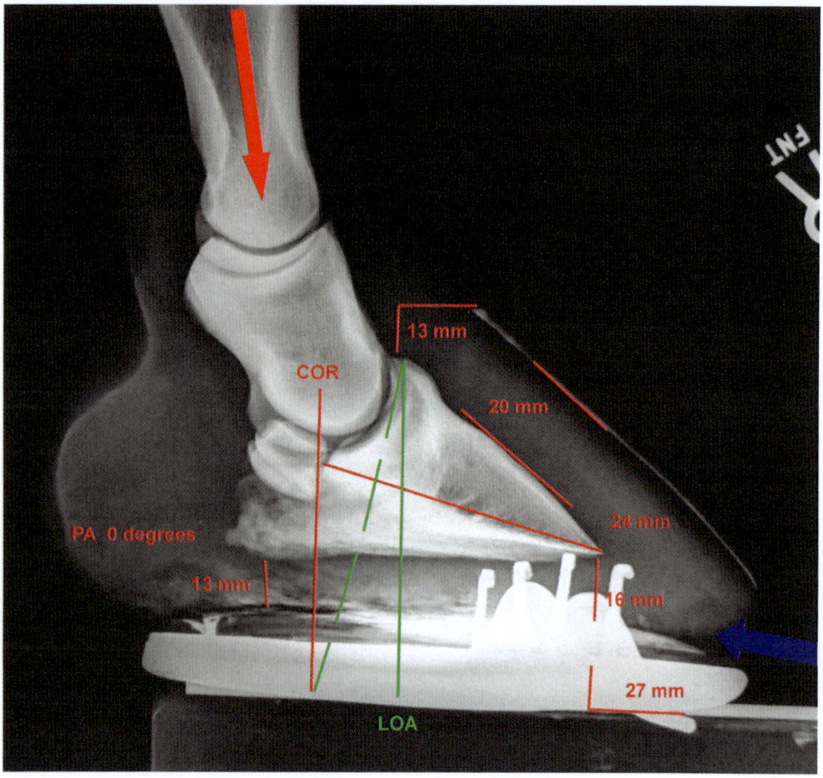

Fig. 11.9: Pre-shoeing radiograph of Case Study 4.

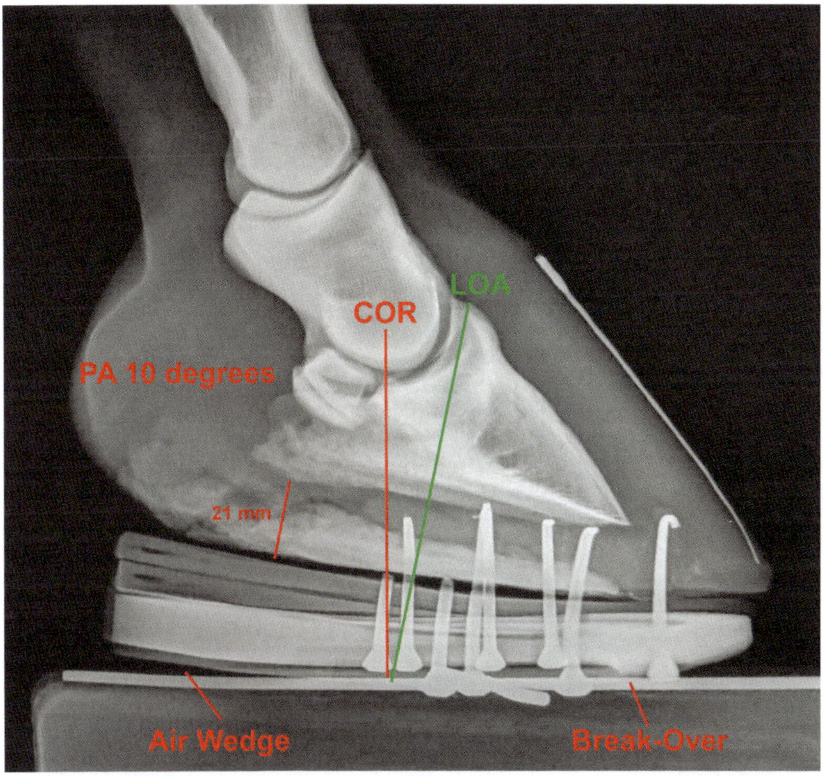

Fig. 11.10: Post shoeing radiograph of Case Study 4; the ventral plane of P3 is parallel to the shoe/pad package. The heels have been reinforced with acrylic to increase ventral depth and protect the existing horn wall from shoe abrasion.

(Case 5)

Case 5: 25% PC: 18 mm; HLp: 21 mm; HLd: 30 mm; SD: 12 mm; HD: 25 mm; CE: 19 mm; BA: 45°; HA: 60°; DBA: 62°; PA 17°; BO: 25 mm

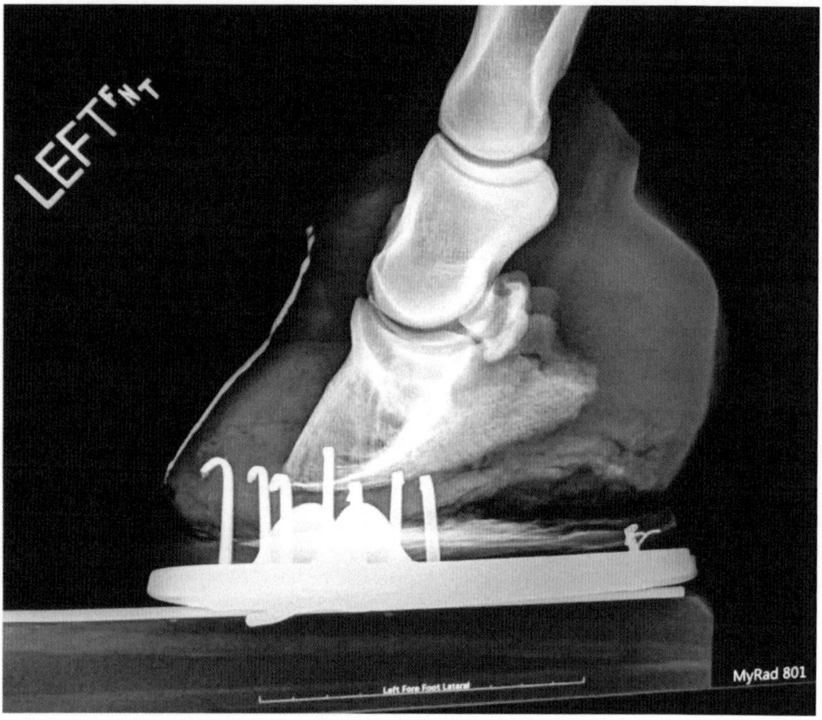

This LF foot (Fig. 11.11) is from the same horse as Case 4 and is a low to medium grade club foot with pathology indicative of a chronic laminitis foot. Because of the remolding of the apex of P3, the palmar cortex of P3 measures 5 mm shorter than that of the RF foot. This should be taken into account when using the PC measurement for soft tissue evaluation, as the remolded bone has shortened the true PC measurement. The cloudy appearance of the HL zone is indicative of a foot with chronic lamella changes or a "Lamella Wedge". Because of the lack of integrity of the dorsal lamella, the foot is displaced distally with a CE of 19 mm and has rotated to a 17° PA in order to equalize tension in the DFT. The BO distance is 25 mm dorsal to the apex of P3 (red arrow): prohibiting cranial rotation, leveraging the dorsal lamella complex, overloading the apex of P3 and compressing the solar corium. SD is 12 mm. The HPA is lined-up into a straight axis from the contracture of the DFT but P3 is not rotated.

Goals:

- Reduce leverage to the dorsal lamella, place BO 20-30 mm palmar to the apex of P3
- Reduce load on solar corium
- Provide mechanics for cranial rotation
- Add ventral depth to foot of 6+ mm

The post-shoeing radiograph (Fig. 11.12) shows the foot reorganized. An acrylic build was added to increase wall length. This was then rockered into a two-plane trim palmarodorsally to the line of action of COR to palmar P3, positioning the wings of P3 in a more parallel plane to the solar surface of the shoe. The radius of the shoe allows adjustment of the PA to 21° as dictated by tension of the DFT. The break-over radius of the shoe is underneath the extensor process (yellow arrow) where the line of action of the GRF at the toe would be, thus relieving leverage to the dorsal lamella.

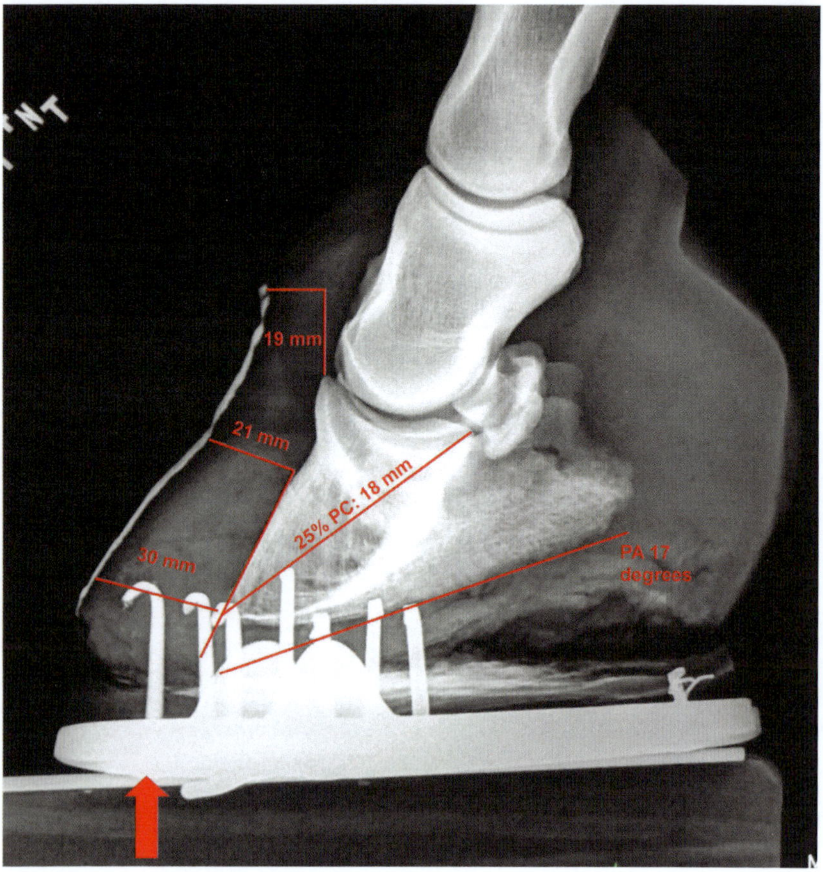

Fig. 11.11: Pre-shoeing radiograph of Case Study 5.

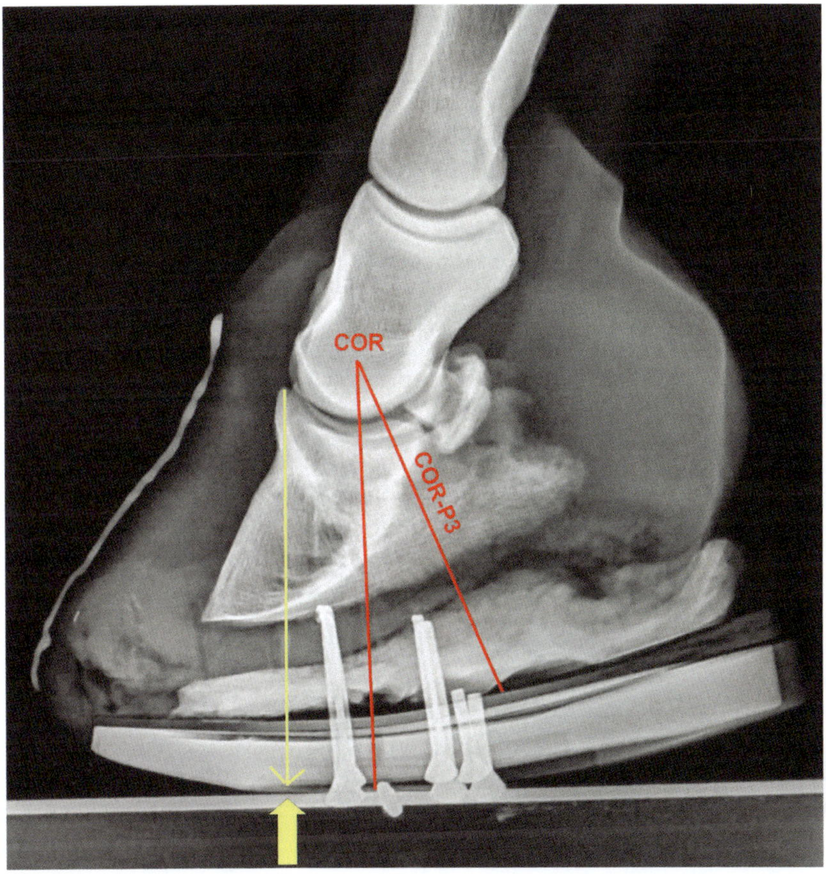

Fig. 11.12: Post-shoeing radiograph of Case Study 5.

(Case 6)

Case 6: 25% PC: 20 mm; HLp: 20 mm; HLd: 16 mm; SD: 25 mm; HD: 27 mm; CE: 4 mm; BA: 44°; HA: 52°; DBA: 47°; PA: 3°; BO: 0 mm

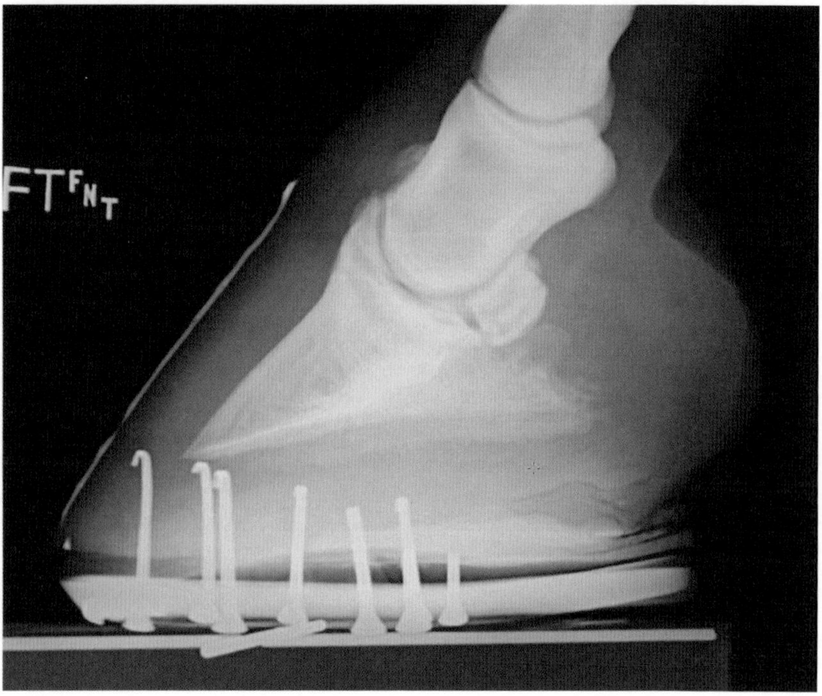

This LF foot (Fig. 11.13) is at six weeks post shoeing. Although this foot is balanced relatively well, the horse is starting to perform poorly. This foot has chronic pathologic changes to proximal P1, distal P2 and on the proximal aspect of the navicular bone. Notice how these changes revolve about COR and that the LOA of the tendons is divergent from COR creating a moment arm to the coffin joint (yellow double arrow). Although the foot sits high in the hoof capsule with a CE of 4 mm, it is caudally displaced 5° between the DBA and HA which gives an inaccurate evaluation to the HPA during physical evaluation. The BO distance is at about 0 mm but the toe is bottomed-out, not allowing further PA increase.

The 25 mm of distal sole offers an opportunity to rocker the toe back to COR which will allow the DFT to cranially rotate the foot.

Goals:

- Trim to reset palmarodorsal and craniocaudal balance
- Slightly increase rocker mechanics

Before shoeing was completed, a 'rocker' shoe was set with two nails (Fig. 11.14) to evaluate mechanics before final nail-up. The radius of the shoe positions BO under the extensor process allowing cranial rotation of the foot into a 10° PA. This converges LOA into COR at the ground surface of the shoe. Before final nail-up, a 0° leather pad and packing was added for frog support. This horse started moving better immediately by optimizing the mechanical ability of the foot.

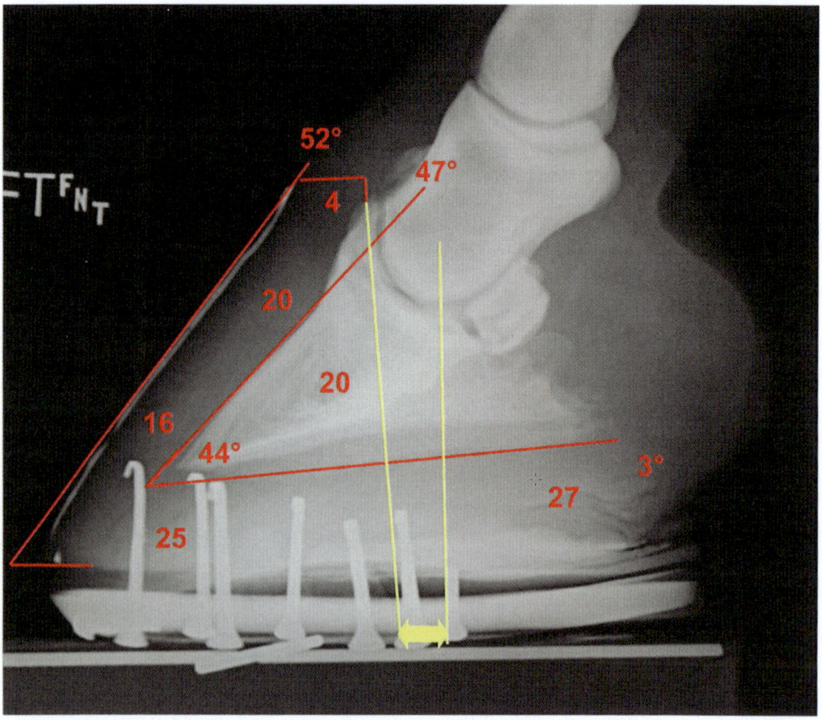

Fig. 11.13: Pre-shoeing radiograph of Case Study 6. This foot has good ventral depth but has lost palmarodorsal mechanics which affect craniocaudal alignment.

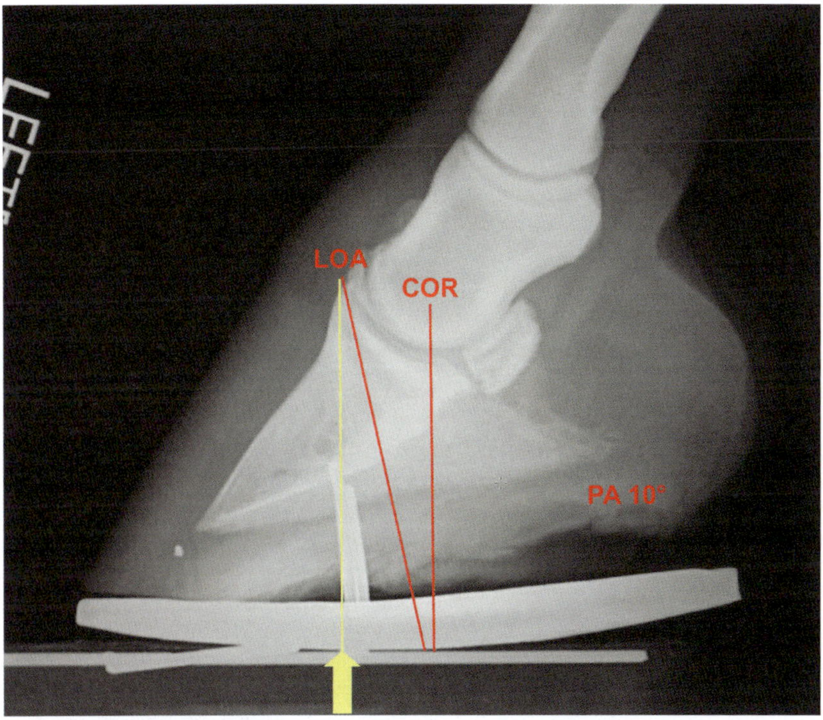

Fig. 11.14: Interim shoeing radiograph of Case Study 6, shoe is set with two nails to check mechanics and balance before final nail-up.

This is the foot in Fig. 7.3 after the application of an acrylic heel build and a rocker shoe with a wedge pad to adjust the palmar angle and align the hoof-pastern axis to a 0° axis in the extended (weight bearing) position. This foot would have a -10° axis in the static (non-weight bearing) position. Pastern extension in the standing horse ranges from 5 - 10°.

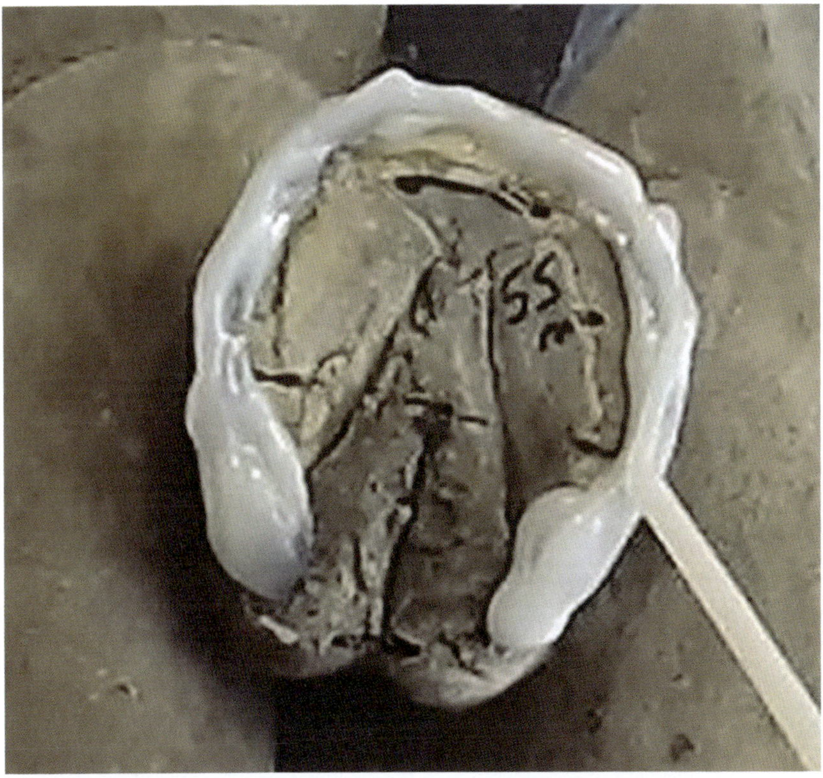

An example of an acrylic wall build to increase ventral depth. The mapping on the foot marks center of rotation (COR) and the dorsal limit of P3 (see Chapter 6).

Glossary

Dorsal Complex: Area of the hoof and lamina dorsal to the extensor process

Goniometer: An instrument for measuring angles

12

EVALUATING THE LAMINITIS FOOT

The laminitis foot can be evaluated as acute or chronic. The acute stage may show very subtle changes that can go unnoticed by the untrained eye. It is at this stage that the application of therapeutic devises are the most effective before major laminar pathology and bone displacement are evident. It is a mistake to let rotation of P3 be the diagnostic criteria for laminitis (Pollitt, 2001). Having the knowledge of the radiographic parameters of a normal foot is the best way of recognizing an acute laminitis case.

The acute phase is when the first signs of foot pain appear. Displacement of P3 can occur in two ways; as a 'rotation' or as a 'sinker'. This depends on the conformation of the foot, tension in the deep flexor tendon (DFT) and the health of the soft tissue relative to center of rotation (COR). The higher palmar angled (PA) foot has more tension in the DFT and will usually have a more robust palmar foot. This foot is more likely to have laminar involvement dorsal to COR and thus rotate (Fig. 12.1). The low PA foot that has a negative dorsal bone angle (DBA) to hoof angle (HA) with weak quarters and digital cushion is more likely to be a 'sinker' with almost complete laminar involvement, these feet will increase PA to some extent as well (Fig. 12.2).

Depending on the severity of the case and treatment, the horse will move into a recovery phase or chronic phase. The recovery phase is a complete grow-off of new hoof-lamina (Fig. 12.3). This may take a year or more. The chronic phase can last indefinitely with mild to severe changes to the lamina and P3 (Fig. 12.4).

The hoof-lamina zone (HL) and coronary band-extensor process (CE) measurements are the first indicators of laminar swelling and distal displacement. If the horse has had radiographs prior to having laminitis, these may be used as baseline markers using the palmar cortex (PC) distance as a guideline. It is important to remember how small the laminar side of the HL zone is. A 16 mm HL zone will have an L component of +-8 mm. A 2 mm increase to this area would correlate to a 25% increase (swelling, tearing) to the lamina. An increase of 2-3 mm is extremely significant (Pollitt 2016).

Other evaluation criteria:

- Extended posture of the pastern indicating a pain contracture
- Loss of sole depth at the apex of P3, radiolucency of the solar corium is not evident
- Appearance of a radiolucent line adjacent to the inter edge of the dorsal wall
- Cranial rotation of P3

Although the mechanical treatment of laminitis is beyond the scope of this book, treatment should be based on recreating equilibrium in the feet for ventral depth, palmarodorsal, mediolateral and craniocaudal balance. Additional mechanics are then added to override the mechanical failure of the foot from laminitis. It must be noted that in severe cases, mechanical therapy may not be able to do this and other options should be explored.

Note: Laminitis cases are complicated and require complete communication between the veterinarian, farrier and horse owner.

Therapy considerations:

- Increasing ventral depth
- Break-over that deleverages the dorsal lamina and allows cranial rotation of the foot
- Reduce leverage to the medial and lateral hoof walls

- Increase PA elevation that equalizes the tension between the DFT and lamina
- Bone support that prohibits distal displacement of P3

Radiographs of the feet post application of therapeutic devices is important as it documents the mechanical goals and gives a baseline for subsequent radiographs (Fig. 12.5). Follow-up radiographs are encouraged every 5 to 10 days to monitor therapy (Redden 1998). Clinical improvement alone can be misleading as some severe cases with a lot of vascular/nerve damage can go through a period of comfort while the foot is numb and undergoing necrosis (Morrison 2004).

At five days post therapy, hopefully the horse is clinically better and the radiographs are stable, meaning no increase in PA, HL, CE or decrease in SD. The posture of P1 may be less extended, indicating a decrease in contractual pain. Some feet may show a measurable increase in sole depth at the apex of P3 from decompression of the solar corium. Continual rotation or sinking is alarming and other options for the feet and horse need to be considered.

At 10 days, there should be a slight improvement with an increase in sole depth of +- 2 mm, a possible decrease in the CE; the PA is stable. Subsequent radiographs should show increased SD and grow-off at the coronet band. As grow-off continues there will be a prominent shelf that delineates the proximal HL and the distal HL zones (Fig. 12.3). The diseased HL zone often becomes wider during grow-off. This is due to its inability to maintain its structural integrity because of the moment arm distance from the toe to COR. In addition to gas lines, the radiograph may reveal a radiolucent spot distal to the apex of P3 from vascular pooling that occurred during the initial insult. This can provide a good marker to determine sole depth during trimming (Fig. 12.6).

A venogram is a technique where a contrast material is injected into the digital vein to visualize the blood flow in the foot (Fig.12.7). The venogram is an effective diagnostic tool if utilized properly. Venograms have detected changes in hoof anatomy as early as 5-7 days after laminitis development (Baldwin, et al 2010).

Laminitis is usually subsequent to an internal condition of the horse or because of support limb compression due to an injury. A radiograph study of the feet of the patient during treatment/surgery can provide information about tissue health and leverages. This information can be used to guide mechanical stabilization of the feet prior to the onset of laminitis.

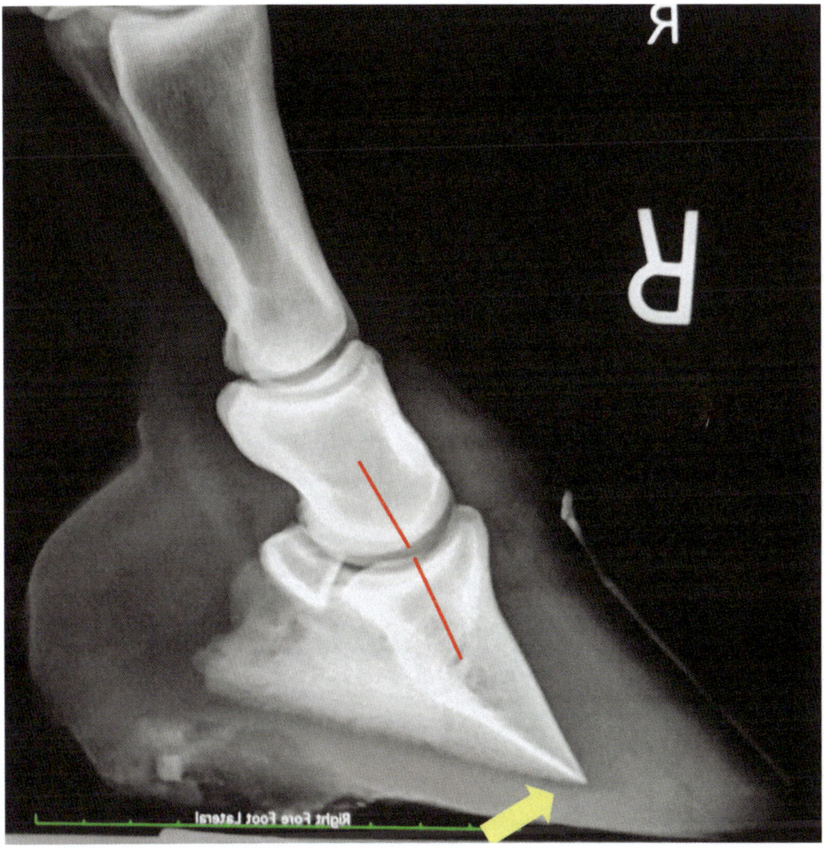

Fig. 12.1: A foot with a 'rotation' type laminitis or "founder". P3 is rotated away from the dorsal wall onto the solar corium (yellow arrow); note alignment of P3 with P2 (broken forward red line).

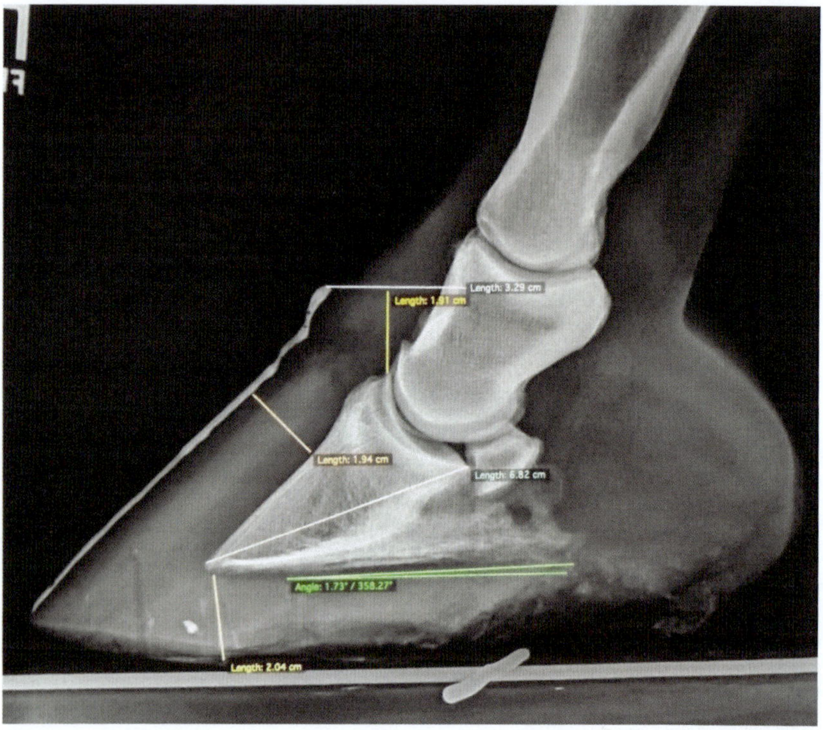

Fig. 12.2: A 'sinker' type laminitis; note the length of the CE measurement at 19 mm, HL zones are 3 mm thicker than normal. Although this foot has ample ventral depth, P3 is compressing the solar corium.

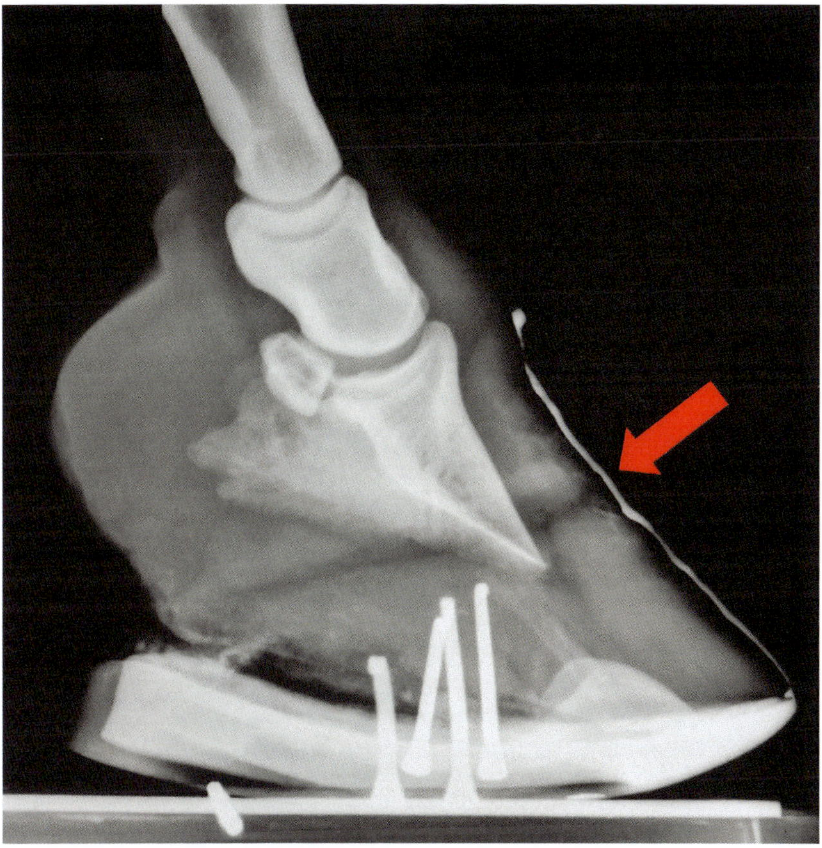

Fig. 12.3: A foot well into the recovery phase. This is the same foot in Fig. 12.1 seven months post-acute laminitis; the red arrow marks the depression in the horn wall when the subcutis of the coronet band was displaced during the initial insult.

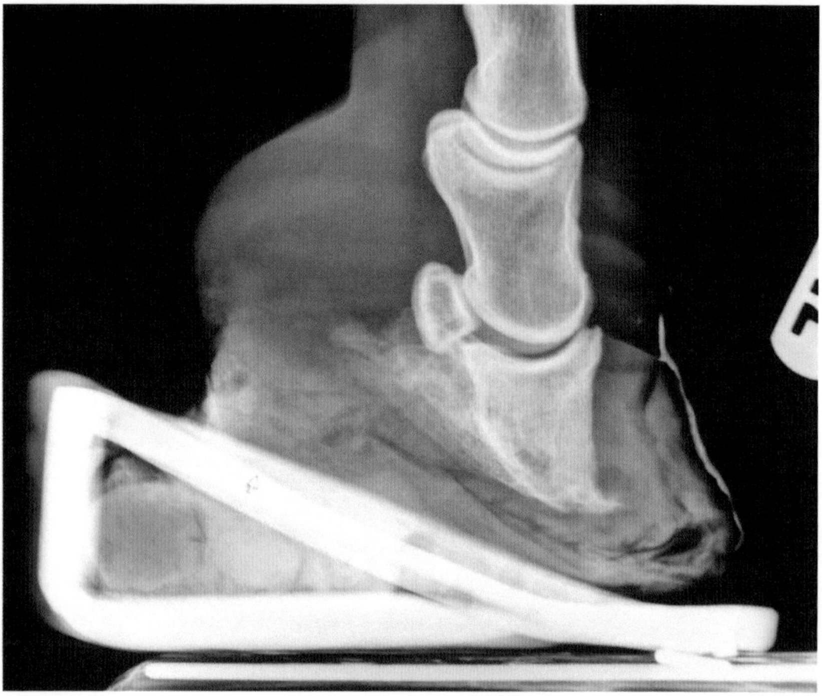

Fig. 12.4: A chronic laminitis foot with major pathology to P3, the dermal lamina and the solar corium. Chronic bone and lamina pain has caused contracture of the deep and superficial flexor tendons.

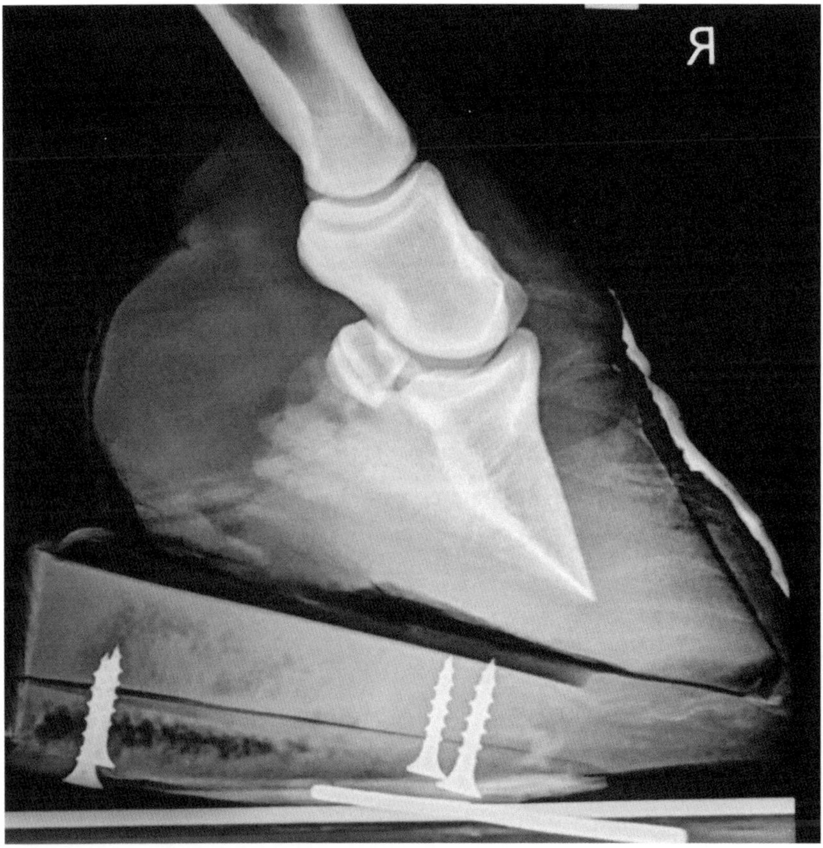

Fig.12.5: This is the foot in Fig. 12.1 post application of a 'Nanric Ultimate'. Measurements of this radiograph would be used for evaluating subsequent radiographs.

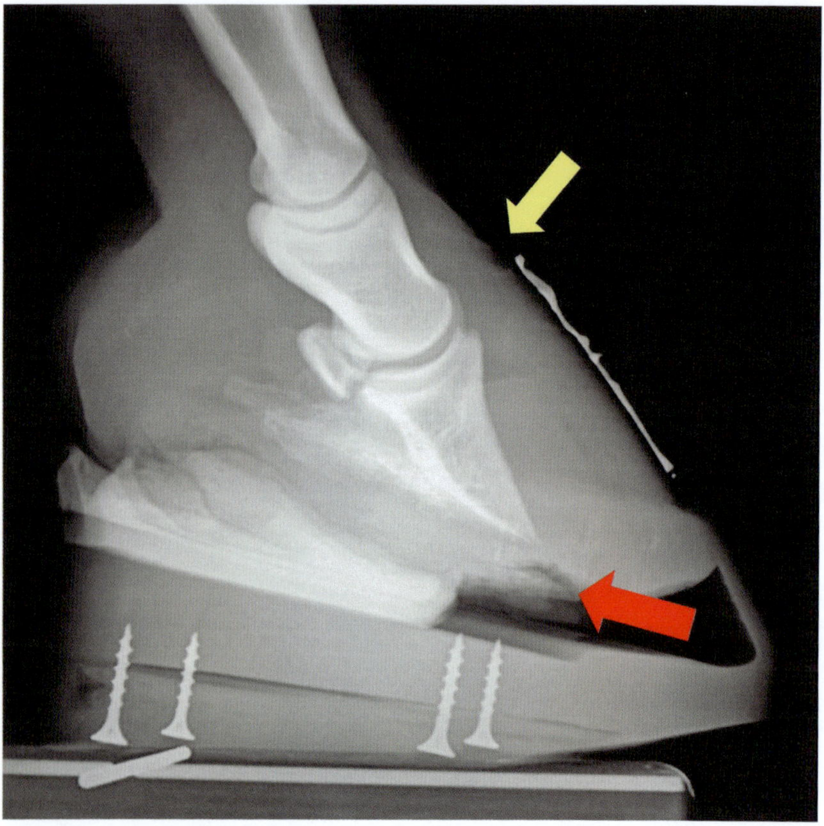

Fig. 12.6: A subsequent radiograph of the foot in Figs. 12.1 and 12.5. The red arrow shows an old area of vascular pooling from compression of P3 onto the circumflex vessels and solar corium. The wings of P3 are staying parallel to the foot surface of the 'Ultimate' and the foot is starting to "grow-off". The radiolucent depression (yellow arrow) at the proximal aspect of the dorsal wall marker is emergence of new horn from the coronet band. There is a gas line in the dermal-epidermal junction of the lamina.

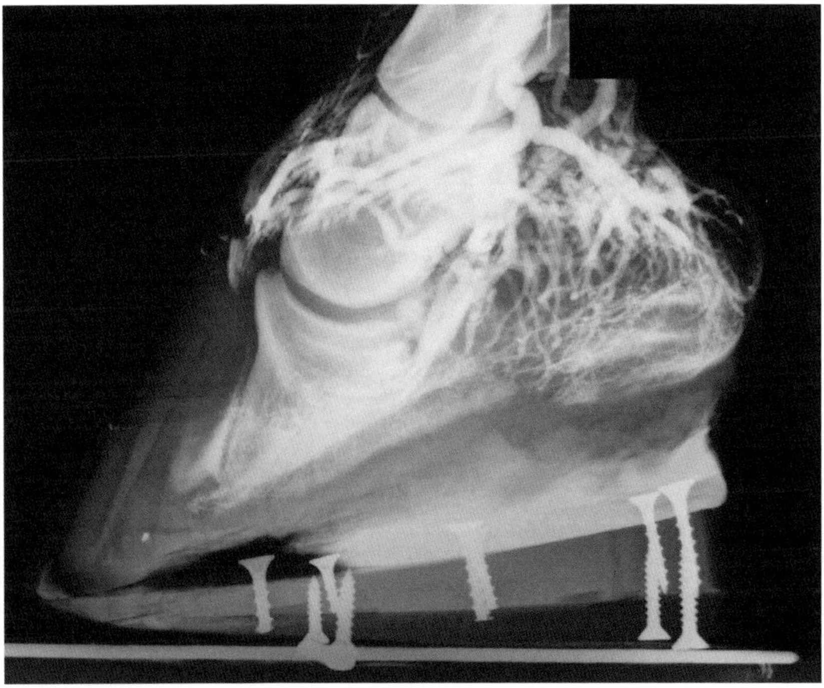

Fig. 12.7: This chronic laminitic case has blood vessels filled with contrast above the coronary band and within the heel. However, within the foot, contrast only fills the terminal arch of the arteries/veins within the coffin bone. Lamellar disease results in compression of the dermis/tissue, creating an absence of contrast in the coronary plexus, proximal Sublamellar bed (in the dorsal wall), and the circumflex and solar vessels (below or distal to the coffin bone) (Rucker A. 2017, personal communication).

Glossary

Acute: Abrupt onset. Connotes an illness that is rapidly progressive, in need of urgent care.

Chronic: Indefinite change, long lasting.

References

Baldwin G.I., Pollitt C.C., Progression of venographic changes after experimentally induced laminitis. Vet Clin North AM. Equine Pract. 2010; 26: 135-140.

Morrison, S., Foot Management, in: Clinical Techniques in Equine Practice, Elsevier 2004, 3:1; 71-82.

Pollitt, C., Laminitis, in: The Illustrated Horse's Foot, A Comprehensive Guide. Elsevier, St. Louis, Missouri, 2016; 177-222.

Pollitt C., Laminitis Radiology in: Equine Laminitis, RIRDC, Kingston ACT 2001; 63-73.

Redden R.F., Treating the Laminitic Horse, in: Understanding Laminitis, The Blood Horse Inc., Lexington KY. 1998: 46-59.

13

EVALUATING RADIOGRAPHS USING MODERN TECHNOLOGY

It wasn't that long ago that the only way to see a radiograph was on film. Film radiographs can offer excellent detail but they have to be developed and are not instantly accessible, also you need a light box to read them.

Digital radiographs are instantly generated and can be seen on the spot. They can be uploaded to a computer where they can be measured or emailed. Radiographs can be printed on plain paper which can be handed out, measured in the field and kept for reference.

There are numerous software programs that allow digital images to be measured on a computer screen using a mouse (Fig. 13.1). A picture of the image can be captured from a cell phone and then sent as a text (Fig. 13.2).

'Metron DVM' (a) is one program that references measurements from calibrated radiograph blocks. The computer software provides step-by-step instructions for inputting and processing the images. A detailed report can then be generated from this information (Fig. 13.3).

Printed images are usually not to scale but can be calibrated using one known component that was measured in whatever program the image was uploaded to. They can also be evaluated using a percentage of the Palmar Cortex of P3 (PC) as it has been established that 25% of the PC should estimate healthy soft tissue areas of the hoof-lamella zone (HL), sole depth (SD) and heel depth (HD) (Fig. 13.4).

Printed images can be measured very easily by hand using a simple protractor or one designed specifically for equine foot radiographs (b) (Fig. 13.5). Bone angles remain constant regardless of image scale.

For prescriptions, images can be downloaded to a computer program which allows graphic mark-ups (Fig. 13.6) or they can be printed out and then marked-up and then scanned and emailed or printed.

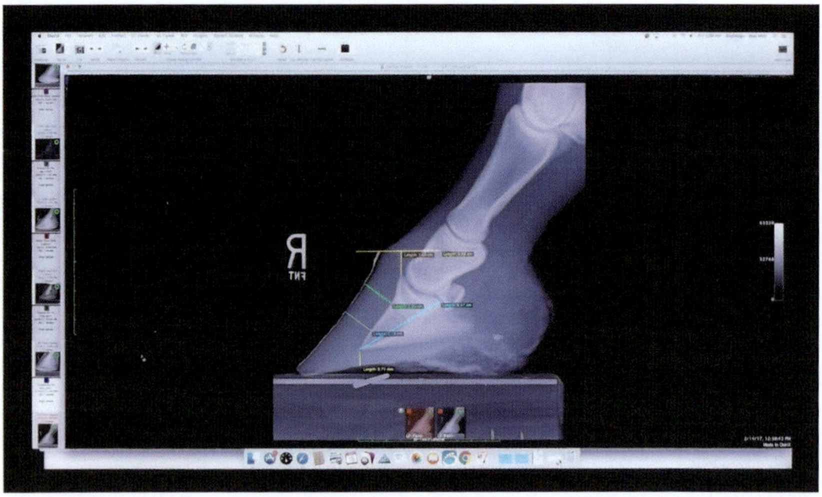

Fig. 13.1: A digital image uploaded to a computer program for evaluation.

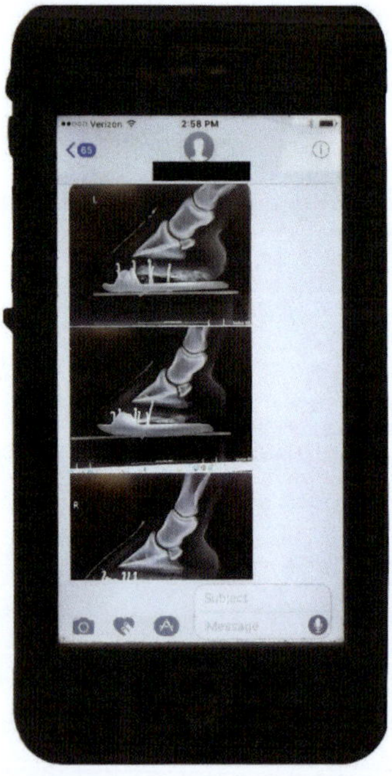

Fig. 13.2: The computer image can be captured on a 'Smart' phone and then sent as a text or images can be emailed.

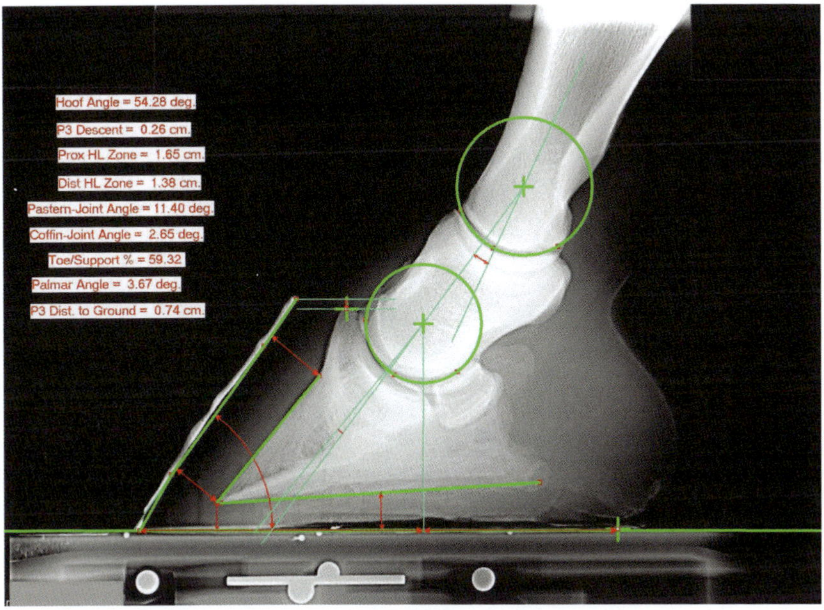

Fig. 13.3: A detailed report can be generated from 'Metron DVM' which uses markers in the blocks for calibration.

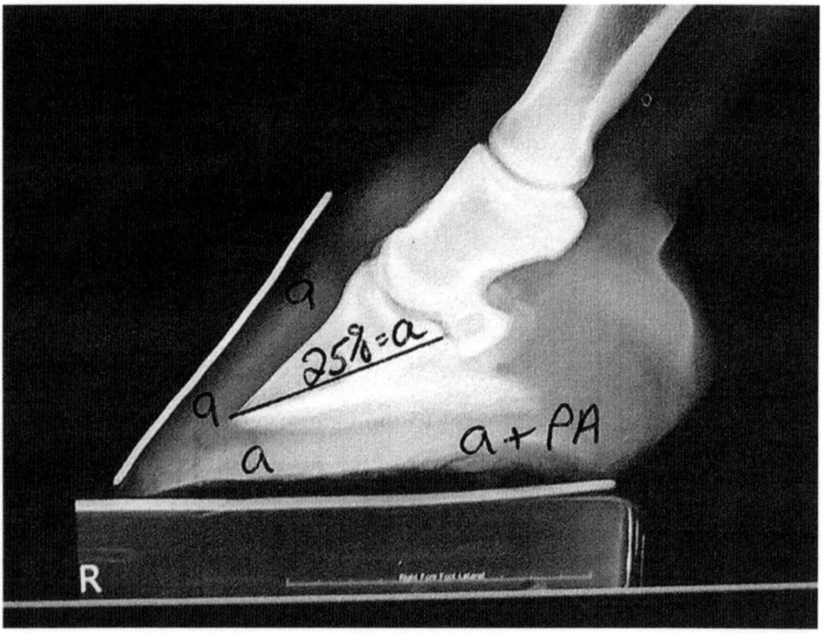

Fig. 13.4: A 4x6 inch print-out of a radiograph; evaluation of soft tissue can be done using 25% of the palmar cortex measurement of P3. Angles stay constant to scale.

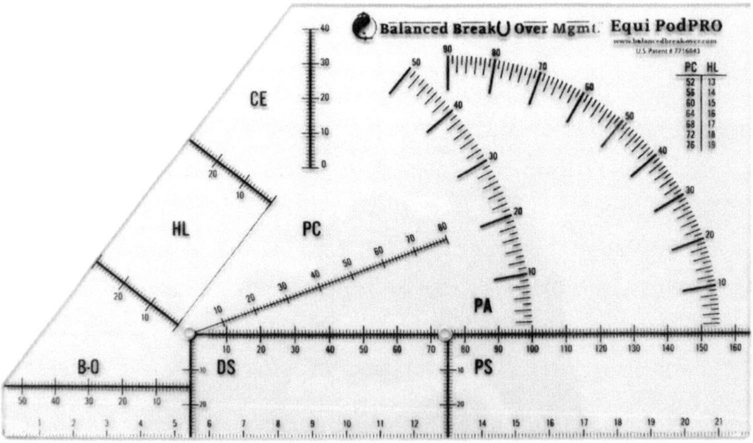

Fig. 13.5: Printed images can be measured by hand very quickly. This protractor by 'Balanced Break-over Management' is designed for lateral radiographs and prompts the user to necessary measurements. A scale in the right-hand corner gives a scale to 25% of the PC measurement.

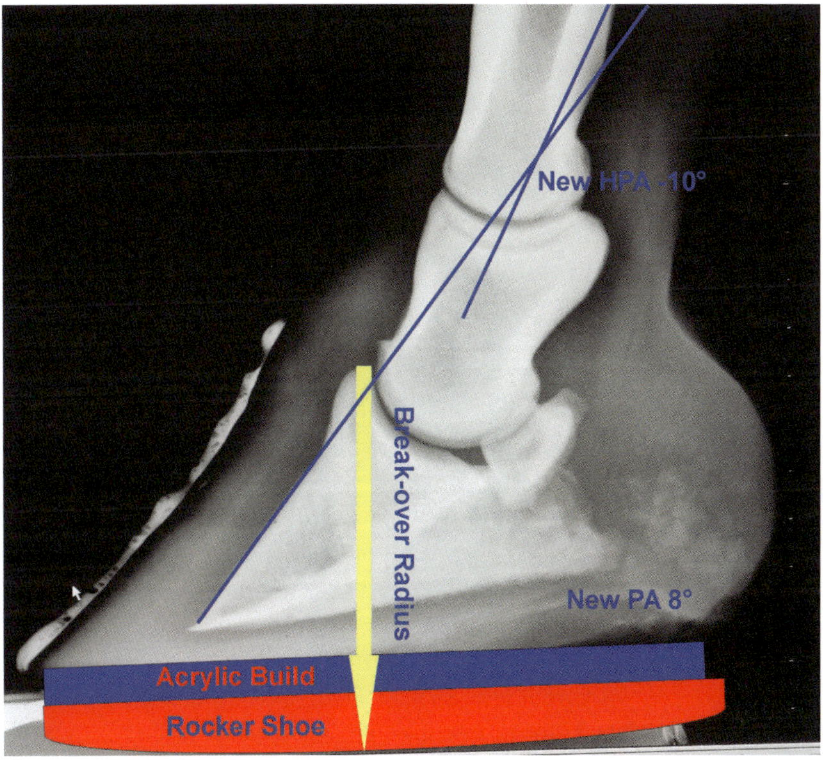

Fig. 13.6: Computer mark-up for a shoeing prescription.

Manufacturers

(a) Metron DVM, Equine Hoof Conformation Software Systems, Epona Tech, P.O. Box 361, Creston, CA 93432.

(b) Equi-Pod-Pro, Balanced Break-Over Management, P.O. Box 704, Los Olivos, CA 93441

A little thought on shoeing prescriptions:

Quite often veterinarians and farriers will give a shoeing prescription in the form of using a particular shoe or pad with no consideration as to how the foot or shoe will be prepared prior to its application.

Instead, be specific about what the horse needs and what goals need to be accomplished in the areas of palmar angle, palmarodorsal mechanics, mediolateral balance and depth of foot. Ask the attending veterinarian or farrier what they can do to accomplish these goals.

There is an old saying that states "There is more than one way to skin a cat than by hanging him from his tail." Hopefully the information in this book will help you accomplish the mechanical needs for the horse.

PETE HEALEY, APFI

Pete Healey learned how to shoe horses as a cowboy in 1979. In 1988 he started a shoeing business and in 2002 he began a working relationship with the renowned Alamo Pintado Equine Medical Center in Los Olivos, CA.

Pete is a graduate of Dr. Reddens' Advanced Equine Podiatry course and holds an APF 1 certificate with the American Association of Professional Farriers. He has been a contributing author for the American Farriers Journal and was a contributing author between 2013 and 2016 for *The Cowboy Way* and *Ranch & Reata* magazines with a series of articles titled, 'Your Horse's Feet'. Pete presents several clinics a year and has lectures around the country.

In 2017, Dr. David Jensen of San Marcos Equine in Los Alamos, CA and Pete founded the Equine Podiatry Education Foundation, a non-profit foundation dedicated to supporting research and education of the equine foot. Pete is continually studying to improve his craft and holds patents to three hoof measuring systems that include a mapping system for the coffin bone and an evaluation system for the hoof-pastern axis.

Pete lives in Santa Ynez, CA with his wife Karen and son Charlie. He is the author of 'The Island, Reminiscences of twentieth century ranching on Santa Rosa Island'.

INDEX